Disaster Ministry Radio Communications Handbook

*Created to assist volunteers
serving through their skills*

First Edition

Gordon L. Gibby MD KX4Z

Volunteers Serving Jesus Through Radio

They were utterly astonished, saying, "**He has done all things
well;** He makes even the deaf to hear and the mute to speak."
Mark 7:37 NAS

An example to emulate in our service....

ISBN: 9781701527607
Independently Published

Note
In order to make distribution of this booklet to volunteers the most efficient, it is provided via Amazon. The price has been set quite modestly, and although I'm not going to make a big spreadsheet out of it, any significant income will be plowed back into ministry or service.

DEDICATION

This introductory text on disaster radio communications, independently written to assist volunteers for the Florida Baptist Disaster Relief group, and volunteers for any similar such group, is dedicated to all the incredible individuals who give of their time, sweat, muscles and treasure to express the Love of Christ in a hurting world.
May your tribe ever increase until
we all meet in Heaven.

Therefore, we are ambassadors for Christ, as though God were making an appeal through us; we beg you on behalf of Christ, be reconciled to God.
2 Corinthians 5:20

INFORMATION ABOUT THE VOLUNTEER
OWNER OF THIS BOOK
Please fill this out.

NAME / CALLSIGN	
HOME ADDRESS:	
HOME PHONE:	
EMERGENCY CONTACT:	
ANY IMPORTANT MEDICAL CONDITIONS	
MEDICATIONS	
ALLERGIES (DESCRIBE REACTION)	
ANY OTHER IMPORTANT DETAILS	

CONTENTS

ACKNOWLEDGMENT

I'd like to acknowledge the tremendous *patience* and *faithfulness*
of my wife,

Nancy E. Gibby

who allows me to perform all kinds of radio experiments, and building campaigns, on the granite kitchen counter-tops, at the breakfast table, and even on the desk in our bedroom -- and that *despite* my have taken up at least half of the 2nd floor for two 24/7/365 RMS stations and an entire bonus room full of old radios, waiting to be repaired "after I get retired."

All who know her, consider her a Saint.
And she even got her Technician License -- KM4YGI.

An excellent wife, who can find?
For her worth is far above jewels.
The heart of her husband trusts in her,
And he will have no lack of gain.
She does him good and not evil
All the days of her life.
She looks for wool and flax
And works with her hands in delight.
She is like merchant ships;
She brings her food from afar.
She rises also while it is still night
And gives food to her household
And portions to her maidens.
She considers a field and buys it;
From her earnings she plants a vineyard.

She girds herself with strength
And makes her arms strong.
She senses that her gain is good;
Her lamp does not go out at night.
She stretches out her hands to the distaff,
And her hands grasp the spindle.
She extends her hand to the poor,
And she stretches out her hands to the needy.
She is not afraid of the snow for her household,
For all her household are clothed with scarlet.
She makes coverings for herself;
Her clothing is fine linen and purple.
Her husband is known in the gates,
When he sits among the elders of the land.
She makes linen garments and sells *them*,
And supplies belts to the tradesmen.
Strength and dignity are her clothing,
And she smiles at the future.
She opens her mouth in wisdom,
And the teaching of kindness is on her tongue.
She looks well to the ways of her household,
And does not eat the bread of idleness.
Her children rise up and bless her;
Her husband *also*, and he praises her, *saying*:
"Many daughters have done nobly,
But you excel them all."
Charm is deceitful and beauty is vain,
But a woman who fears the Lord, she shall be praised.
Give her the product of her hands,
And let her works praise her in the gates.

Proverbs 31: 10-31

1 INTRODUCTION

Disasters disrupt virtually all facets of normal, organized society. Utilities such as potable water and electrical power may be compromised or severely damaged. Food and fuel delivery may be interrupted. **But one of the most devastating effects of a disaster is to damage communications, because the efficiency of modern societies depends on communications.** [1]

Societies are now accustomed to instantaneous and reliable communications.

- Sub-units of almost all organizations (government, NGO, businesses, hospitals, and even families) are directed and monitored by Internet data communications or cellphones.
- Equipment and supplies are ordered by computer, Internet and telephone.
- Appointments and schedules are created, organized, and promulgated by web pages, Facebook and online entry systems
- Questions can be answered almost immediately via cell phone or text instant messaging.
- Electrical power, liquid and gaseous fuel deliver, sewage, and water systems are maintained and tracked by automated systems that depend on fiber optic, radio, or copper-based electronic

---

[1] Strongly suggest you read this excerpt: https://www.qsl.net/nf4rc/KatrinaComms.pdf to understand just how devastating it can be to lose communications.

connections.

When a disaster damages or obliterates communications systems, not only are food, water, shelter and protection no longer a "given," but the entire community, as well as responders, suddenly lack the ability to operate with their normal efficiency, because of a lack of normal organizational communications. The Florida Baptist Disaster Relief arrived at Mexico Beach after Hurricane Michael to find *everything* obliterated -- and they had NO communications -- had to send cars out by a very difficult and slow trek to carry messages. As a result, they became much more interested in emergency backup communications, and hence this book took shape. **It becomes difficult to even know where the problems are, and what is their magnitude**, much less to coordinate an effective, efficient response, when communications are no longer available.

A very practical and concrete local example might be: *A mudding team is heading out in a disaster area to work on three houses. Communications in the area have been destroyed and the team has no means of communications. They are unable to find one address. With no communications, they are unable to check the address or their directions. When they reach the next address, it has already been attended to -- but there were no communications of that. On the way to the third address, they suffer significant vehicle difficulties and become stranded -- and with no way to communicate their plight to "home base" now some 10 miles (3 hours of walking) away. Night is approaching and they have gotten nothing done, and are now stranded in a disaster area. Thankfully a sheriff's deputy drives past with an amateur radio volunteer in the vehicle who has working communications -- they had been alerted over amateur radio that the mudding team was overdue and had been searching for them.*

The purpose of this booklet is to provide **technical education** to volunteers for Florida Baptist Disaster Relief amateur radio communications group, in order to help them better serve their organization in service to others.[2] The focus of this booklet is more on

---

2   Nehemiah is one of my heroes His advance planning became obvious in his detailed response to the King's question: "Then the king said to me, 'What would you request?' So I prayed to the God of heaven. I said to the king, 'If it please the king, and if your servant has found favor before you, send me to Judah, to the city of my fathers' tombs, that I may rebuild it.'

the technical aspects, than the organizational aspects. In the Incident Command System, there is already a hierarchy that provides leadership and direction of the entire organization to accomplish its goals, and educational materials on the ICS system are readily available. This document will focus more specifically on the **skills, assets and strategies** of the Communications Unit and its volunteers, to provide flexible solutions in the service of the ICS organization.

You, the volunteer for Communications, should make every effort to take the basic ICS courses listed below, and fill in the date of your completion of them (an often-requested bit of information when listing your credentials). I can tell you from personal experience, having done a few dozen of these courses now, that it was IS-300 where the "light went on" for me and I began to realize how powerful is this system, bringing together hundreds to thousands of volunteers who have never worked with each other, and instantly melding them into a cohesive organizational structure with many pre-made job titles and responsibilities. The ICS system can be utilized for just about any task. Our ham group has used it for tower antenna work. It provides structure for all the information a group needs to accomplish a mission.

IS-100x   _____

IS-200x   _____

IS-700x   _____

IS-800x   _____

IS-120x   _____

ARRL EC-001   _____

For Advanced Training:

IS-300   _____

IS-400   _____

Then the king said to me, the queen sitting beside him, 'How long will your journey be, and when will you return?' So it pleased the king to send me, and I gave him a definite time. And I said to the king, 'If it please the king, let letters be given me for the governors _of the provinces_ beyond the River, that they may allow me to pass through until I come to Judah, and a letter to Asaph the keeper of the king's forest, that he may give me timber to make beams for the gates of the fortress which is by the temple, for the wall of the city and for the house to which I will go.' And the king granted _them_ to me because the good hand of my God _was_ on me. " Nehemiah 2: 4-8

IS-139 _____

The Professional Development Series is also quite useful.

2 FLORIDA BAPTIST SPECIFIC ICS INFORMATION

(This material is my best understanding of their methods and
needs; consult with local officials and volunteers for additional
or improved information.
Your specific organization may be similar.)

This drawing is loosely based on information from Missouri
Baptist Disaster Relief, but likely applies well in many states:

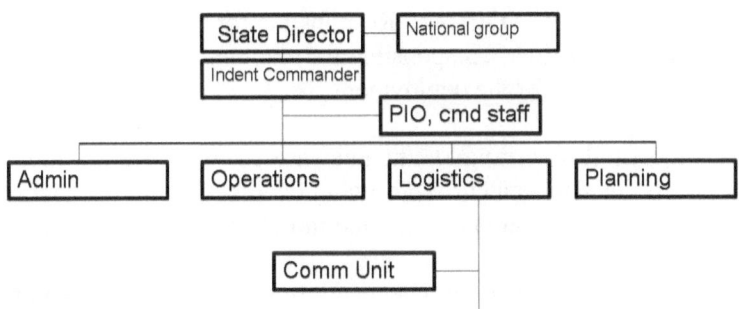

Illustration of ICS System in disaster relief. G. Gibby

Key concept of ICS: you "stand up" additional positions, and
units, *as they become necessary*. If the Incident Commander is
the only person -- by definition they fulfill every responsibility.
As additional ICS staff are appointed, tasks become further
divided.

Key Officers in any ICS-based response include:

- The Incident Commander -- in charge of all the units.

COMMAND STAFF

- Safety Officer: Reports to the Incident Commander; identifies and deals with unsafe conditions. Oversees all health and safety aspects of the volunteers

- Public Information Officer (PIO) Reports to the Incident Commander. Handles all statements to the public, media and other agencies.

GENERAL STAFF

- Operations Officer: responsible for the recovery units
- Logistics Officer: responsible for housing, food, general supplies and trailer/vehicle parking (Communications comes under Logistics in current ICS documents.)
- Administrations Officer: Responsible for reports, paper work.
- Planning Officer: responsible for planning including preparations for the deployment.

At the time of this writing, I am unaware of any specific additional requirements for DR amateur radio volunteers, beyond those required of all volunteers: background checks, and initial training in some area.

We would like to see all amateur radio communications volunteers develop the following capabilities as they mature in their skills:

- General Class or higher license (allows longer-range communications; Technicians can operate with higher level supervision)
- Ability to read and understand Incident Action Plans, including ICS-205 / 205A
- Ability to move formal and informal traffic by voice (SSB or FM)
- Ability to move formal traffic via WINLINK (optionally, via additional systems)
- Capability of handling communications tasks using mobile or handheld business band radios
- Understanding of antenna and power systems necessary to provide adequate communications in the field.
- Understanding of the satellite / VOIP / associated systems

3 OVERVIEW OF COMMUNICATIONS GOALS

This graphic gives a good overview of the types of communications you may be assisting in a Disaster Relied Ministry:

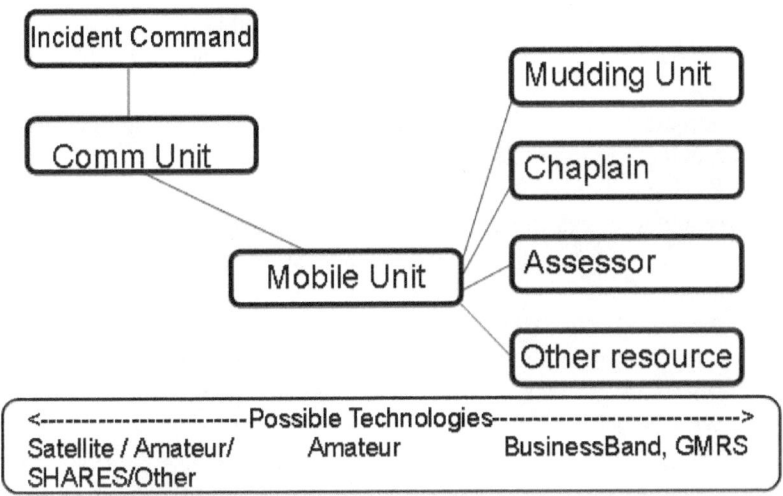

Possible Radio Technologies. Illustration by G. Gibby

In any disaster area of operations with damaged communications, multiple types of communications links must be re-created, including:

- Between Disaster Relief Incident Command (IC) and State or Local Emergency Operations Center(s) and county / state / federal officials at their locations, potentially through volunteers from the Disaster Relief stationed there.

- Between DR IC and suppliers of all types, including materials for food preparation and all other needed supplies

- Between DR IC and forward deployed units.

- Between DR IC / forward deployed units, and strike teams or individual resources such as mudding teams, chaplains etc. in the field.

Some of the communications are long range, some are short range. The available radios, frequencies, modulations and systems to make all this work may be quite diverse. Examples include:

- Amateur radio HF voice nets to communicate with State EOC, or other DR teams
- Amateur radio HF digital systems (either broadcast or WINLINK) to deliver detailed information and/or hard copy to suppliers, media, county/state/federal officials
- SHARES HF voice or digital systems to accomplish similar goals.
- Amateur radio VHF/UHF simplex or repeater communications to reach deployed units or individual resources
- Business band VHF/UHF simplex or repeater communications to reach deployed units or individual resources
- GMRS or other suitable radio communications to respond to the needs of citizens
- Amateur radio HF digital systems (principally WINLINK) to relay scores to thousands of messages from survivors outbound from the disaster area to concerned loved ones ("Health and Welfare")[3]
- VSAT satellite communications with internet-like switching systems to individual computers in a command locations
- Human relaying of messages from one system to another
- VOIP (voice over Internet Protocol) telephone systems
- Desktop and laptop computers
- Temporary ("itinerant") repeater systems to assist in communications.
- Appropriate communications with COM-L and other ICS personnel managing communications for county/state/federal or other NGO organizations.

---

3   Because survivors are often not at their normal locations and not reachable through normal communications, INBOUND health and welfare traffic to a disaster area can be very difficult to deliver and is discouraged, particularly early in a disaster.

HOW TO MAKE CONTACT WHEN NORMAL STUFF IS BUSTED OR OVERWHELMED IN A DISASTER SITUATION

DISTANT COMMUNICATIONS

Voice Tactical & Logistical Resources: ARES ICS-205

The Florida Sections are very likely to release an ICS-205 giving expected HF and possibly other frequencies that will be in use for ARES-ARRL communications -- and **these will give you quick voice connections to help of all sorts**. Explain your position representing an NGO (Florida Baptist Disaster Relief or equivalent) and they are very likely to help you with any communications you need.

Here is Page One of an actual **ICS-205** utilized by ARES in Florida during the run-up to Hurricane Dorian (2019):

INCIDENT RADIO COMMUNICATIONS PLAN (ICS 205)

1. Incident Name: Hurricane Dorian			2. Date/Time Prepared: Date: 9/3/19 Time:				3. Operational Period: Date From: 9/3/19 Date To: 9/4/19 Time From: 0800 Time To: 0800			
4. Basic Radio Channel Use:										
Zone Grp.	Ch #	Function	Channel Name/Trunked Radio System Talkgroup	Assignment	RX Freq N or W	RX Tone/NAC	TX Freq N or W	TX Tone/NAC	Mode (A, D, or M)	Remarks
		Command	FARES Net		3960 LSB		3960 LSB		A	Florida ARES Net
		Command	FARES Net Alt		7242 LSB		7242 LSB		A	Florida ARES Net Alternate Frequency
		Command	SARNET		See		Attach		A	Statewide Amateur Radio Network
		Interoperability	National Calling Frequency		146.52		146.52		A	National Calling Interoperability Frequencies
		Interoperability	National Calling Frequency		446.00		446.00		A	National Calling Interoperability Frequencies
		Interoperability	National Calling Frequency		144.39		144.39		D	National APRS Frequencies
5. Special Instructions: All HF frequencies are PLUS OR MINUS five khz depending on conditions and interference. SARNET. See Attached frequency List.										
6. Prepared by (Communications Unit Leader): Name: Karl Martin K4HBN SEC Signature:										
ICS 205		IAP Page 1/1	Date/Time: 9/2/19							

ICS-205 from Florida ARES(R) Hurricane Dorian

Note that this document gives you important frequencies where you can find nets and HELP in the disaster theater.

It would be a great idea to have in your preparations the email addresses for the CURRENT Section Emergency Coordinators for the disaster area, so that you can ask about an ICS-205. (You can use your WINLINK skills to get out any email you need -- as long as you have an address!)

The COMM LEADER for the DISASTER RELIEF group should certainly be gaining this information and then **creating an ICS-205 specific for the DISASTER RELIEF effort --- and then forwarding that document to the COM-L in charge of the county/state/federal response effort**. That can be extremely helpful information, and will likely get you "in the loop" for information updates from the authorities.
4

FLORIDA SARNET

Florida has a unique system of connected UHF repeaters that runs the backbone of the state, connected together in a giant "Party Line" system. The repeaters are connected by radio equipment that receives and transmits to local ham UHF repeaters in many communities, but is connected together up and down the Interstates via microwave connections formerly used by the Florida Motorist Assistance System (which was decommissioned after cell phones became so widespread).

The SARNET offers party-line ANALOG FM VOICE (NOT DIGITAL) communications and will likely be monitored by the State EOC as well as EOC's in many counties. The SARNET has many strengths and many weaknesses, but it is an important tool in your bag of capabilities:

- Allows easy voice connection to a host of authorities in a time of real need

- Can be reached with just a UHF transceiver; no computer or special knowledge needed

- Because of the delays, you'll need to be even more patient -- after hitting the PushToTalk button, your first second or two will

---

4   Lt. Rulapaugh tells me that when he deployed the Region 3 MARC unit to the panhandle and submitted his ICS-205 for his team's operations, he was rewarded with a steady stream of UPDATES, since now authorities knew how to reach him.

not be heard --- so give it a moment for all the relays to click in....

• This system is a "serial" system, and any failure of the microwave connecting backbone splits it into broken pieces of spaghetti. That can easily happen in a severe disaster.

• This system is susceptible to malicious jamming. While there are some techniques that can be utilized against jammers, they aren't perfect. So it is a vulnerable resource.

• DON'T HOG THE FREQUENCY -- this is used by many, many people, so be BRIEF and SUCCINCT. If its operating as a "directed net" don't say anything until the Net Control Station requests input.

You need to keep with you a list of all the SARNET outlets. Here is an image of Page 2 of the Florida ICS-205 (Amateur Radio) for Hurricane Dorian:

SITE	Repeater Receive	RX Tone	Repeater Trasmit	TX Tone
Apalachiola	449.4000	94.8	444.4000	94.8
Brooksville	449.8250	146.2	444.8250	146.2
Central Turnpike	449.9750	107.2	444.9750	107.2
Chattahoochee	449.9750	94.8	444.9750	94.8
Chipley	449.7500	100	444.7500	100
Clermont	449.9750	103.5	444.9750	103.5
Cocoa	449.6500	107.2	444.6500	107.2
Crestview	449.9000	100	444.9000	100
Dundee	449.3500	103.5	444.3500	103.5
Florida City	447.0500	114.8	442.0500	114.8
Ft. Lauderdale	447.8500	110.9	442.8500	110.9
Ft. Myers	449.2250	136.5	444.2250	136.5
Gainesville	449.9250	123	444.9250	123
Islamorada	447.8500	114.8	442.8500	114.8
Jacksonville	449.7000	127.3	444.7000	127.3
Key West	449.4000	114.8	444.4000	114.8
Lake City	449.9000	110.9	444.9000	110.9
Lakeland	447.2750	82.5	442.2750	82.5
Live Oak	448.7000	110.9	443.7000	110.9
Madison	449.3000	94.8	444.3000	94.8
Miami	449.6000	167.9	444.6000	167.9
Milton	449.7250	100	444.7250	100
Naples	449.9500	103.5	444.9500	103.5
Ocala	449.0250	123	444.0250	123
Orlando	449.0750	103.5	444.0750	103.5
Palm Beach	448.9750	110.9	443.9750	110.9
Panama City	449.1750	100	444.1750	100
Pensacola	449.8750	100	444.8750	100
Perry	448.1000	94.8	443.1000	94.8
Sarasota	449.8000	100	444.8000	100
Sebastian	449.3750	107.2	444.3750	107.2
SEOC	448.5000	94.8	443.5000	94.8
Skyway Bridge	447.2500	146.2	442.2500	146.2
St. Augustine	447.8000	127.3	442.8000	127.3
Stuart	449.1500	107.2	444.1500	107.2
Tallahassee	447.1000	94.8	442.1000	94.8
Tampa	447.8500	146.2	442.8500	146.2
Yulee	447.9000	127.3	442.9000	127.3

Updated on 8/2/19

Further information about the SARNET:

Homepage: https://www.sarnetfl.com/

How It Works: https://www.sarnetfl.com/how-it-works.html

SARNET CURRENT MAPS: https://www.sarnetfl.com/system-maps.html

Page to check current status: https://www.sarnetfl.com/system-status.html

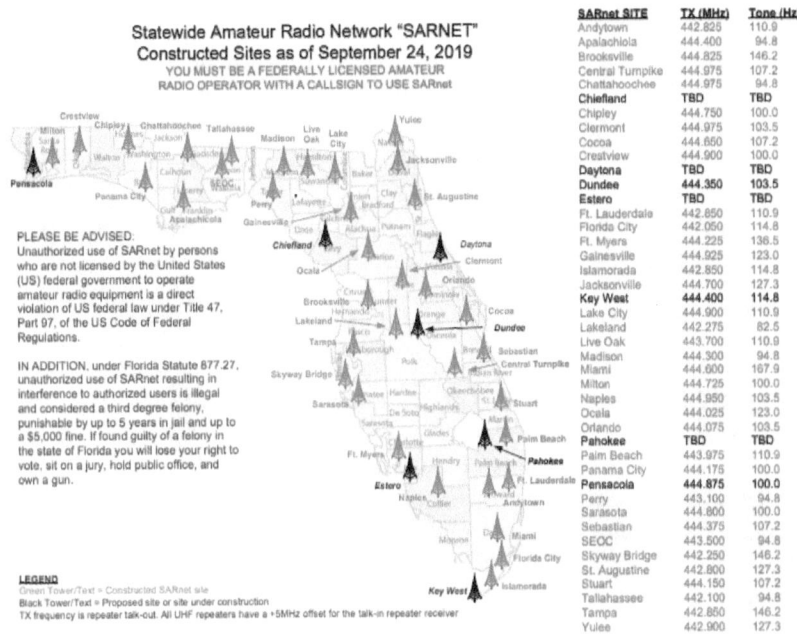

Statewide Amateur Radio Network "SARNET"
Constructed Sites as of September 24, 2019
YOU MUST BE A FEDERALLY LICENSED AMATEUR
RADIO OPERATOR WITH A CALLSIGN TO USE SARnet

SARnet SITE	TX (MHz)	Tone (Hz)
Andytown	442.825	110.9
Apalachicola	444.400	94.8
Brooksville	444.825	146.2
Central Tumpike	444.975	107.2
Chattahoochee	444.975	94.8
Chiefland	**TBD**	**TBD**
Chipley	444.750	100.0
Clermont	444.975	103.5
Cocoa	444.650	107.2
Crestview	444.900	100.0
Daytona	**TBD**	**TBD**
Dundee	**444.350**	**103.5**
Estero	**TBD**	**TBD**
Ft. Lauderdale	442.650	110.9
Florida City	442.050	114.8
Ft. Myers	444.225	136.5
Gainesville	444.925	123.0
Islamorada	442.850	114.8
Jacksonville	444.700	127.3
Key West	**444.400**	**114.8**
Lake City	444.900	110.9
Lakeland	442.275	82.5
Live Oak	443.700	110.9
Madison	444.300	94.8
Miami	444.600	167.9
Milton	444.725	100.0
Naples	444.950	103.5
Ocala	444.025	123.0
Orlando	444.075	103.5
Pahokee	**TBD**	**TBD**
Palm Beach	443.975	110.9
Panama City	444.175	100.0
Pensacola	**444.875**	**100.0**
Perry	443.100	94.8
Sarasota	444.800	100.0
Sebastian	444.375	107.2
SEOC	443.500	94.8
Skyway Bridge	442.250	146.2
St. Augustine	442.800	127.3
Stuart	444.150	107.2
Tallahassee	442.100	94.8
Tampa	442.850	146.2
Yulee	442.900	127.3

PLEASE BE ADVISED:
Unauthorized use of SARnet by persons who are not licensed by the United States (US) federal government to operate amateur radio equipment is a direct violation of US federal law under Title 47, Part 97, of the US Code of Federal Regulations.

IN ADDITION, under Florida Statute 877.27, unauthorized use of SARnet resulting in interference to authorized users is illegal and considered a third degree felony, punishable by up to 5 years in jail and up to a $5,000 fine. If found guilty of a felony in the state of Florida you will lose your right to vote, sit on a jury, hold public office, and own a gun.

LEGEND
Green Tower/Text = Constructed SARnet site
Black Tower/Text = Proposed site or site under construction
TX frequency is repeater talk-out. All UHF repeaters have a +5MHz offset for the talk-in repeater receiver

Note that this map is only current as of October 7 2019

RADIO EMAIL: WINLINK
(AMATEUR OR SHARES)

Your primary tool for sending long lists or heavily detailed information is very likely going to be WINLINK. This *best-in-class* amateur radio system for handling email and attachments has a long history of serving in disaster situations as well as in normal everyday communications. There are upwards of 20,000 registered users. Somewhere around 40-50,000 messages are transferred through this system every month. There are approximately 100 HF "Radio Message Servers" (RMS) stations provided by VOLUNTEERS throughout the United States, and more around the world.[5] It is the premier system as

---

5   There are a huge number of additional VHF/UHF short-range Gateways, but these are not generally useful for long range messaging in a disaster.

of this date, for radio data traffic handling. A later chapter in this document will give you more information.

The WINLINK system allows the use of multiple different digital techniques, including:

- PACTOR (requires an expensive hardware modem)
- WINMOR -- uses software and a soundcard, built-in to WINLINK software
- ARDOP -- uses software and a soundcard, built-in to WINLINK software just like WINMOR
- VARA -- 3rd party soundcard software, requires a small fee for fast operation

If other communications fail, you are very likely to be using WINLINK for ordering supplies and all communications that need to be officially recognized. *Remember that a voice tactical message can be easily lost or ignored -- a written communication using WINLINK or another data system can create a "paper trail" where justifications are more easily recognized or obtained, and performance is more guaranteed.* This was one of the big lessons ARES(R) and Florida Baptist Disaster Relief learned in Hurricane Michael.

----------------EXAMPLE OF FOOD ORDER-------------------

Date: Tue, Oct 16, 2018 at 4:38 PM
Subject: Hiland Park
To: Sue Smith USF <madeupname@usfoods.com>

Please add 4 cases 10 ounce cups.
Case Poly one size fits all
gloves.
4 cases Cam liners/garbage bags
Pallet Gatorade
Bread for sandwiches
Sandwich meat, cheese
Sandwich bread for 100 for 3 days
No breakfast food.
Peanut butter and jelly

Quat sanitation tablets.1
Quat sanitation test strips.
Snacks
Zip lock type sandwich bags
Brown paper sandwich bags
Fruit cups or fruit to match number entrees.
Prefer canned, but all okay as long as not raw.
Thaw and serve bread please.

Thanks for your patience

=======================================

That kind of written information is EASY to transmit using WINLINK in less than a couple minutes, **if you have already discovered which stations are strongest on which bands, at which hours.**

WARNING: Do NOT send orders for items to firms <u>outside of a real disaster</u> on *amateur* frequencies as non-disaster commercial communications are FORBIDDEN in amateur radio.
OK to use SHARES for training like this.

SUGGESTIONS TO AVOID REGULATORY DIFFICULTIES WITH AMATEUR RADIO AND NON-AMATEUR BUSINESSES IN DISASTER COMMUNICATIONS

Explanation: There are persons who will take issue with any communications related to purchase of items over amateur radio, particularly if these accidentally occur outside of a real disaster.

SUGGESTION 1: Add in the phrase
"EMCOMM Traffic in support of Florida Baptist Disaster Relief"
-- somewhere in the subject line or the Body of the message sent over amateur frequencies to make it more clear in a disaster amateur radio communication.

SUGGESTION 2: You can send these kinds of "ordering supplies" messages **through the SHARES Winlink system** and avoid all of the "pecuniary interest" issues..... It works exactly the same as the ham Winlink system -- but requires your SHARES license to be selected, and uses federal frequencies outside the Amateur bands. ONLY PACTOR -- no sound card modes for that version of Winlink.

SUGGESTION 3: After the disaster is over, on a computer go to winlink.org and log into your *amateur radio callsign* user account:

https://winlink.org/user

then click on "My Whitelist" left side of page (https://winlink.org/content/whitelist_manager) and DELETE any businesses with which you had communications during the disaster. This will prevent them from inadvertently sending you email outside of the disaster. Notify businesses of normal email addresses for usage outside of disasters.

You may also be sending out many other types of information via WINLINK (either amateur or SHARES), including:

- email letters from local admin to higher-up admin in Florida Baptist
- simple spreadsheets (possibly reduced to `.txt or .csv` format to save space)
- lists of volunteers
- hours lists
- important communications to representatives at State or Local EOCs
- anything else that is important to keep missions going and volunteers safe

OTHER LONG RANGE SYSTEMS

National Traffic System / Radio Relay International
Traditional voice, cw or other traffic nets may well be available for your usage -- capture lists of their frequencies before you go, or get them from your COMM LEADERS

National Traffic System - DIGITAL
Using older WINLINK systems, this is a hybrid system that can take digital input, and automatically move it to human-powered traffic nets.

SHARES: Federal Voice & Digital Resource
The Shared High Frequency Resources system of the Federal Government uses NTIA-delivered licenses that let authorities and significant NGO's have their own HF communications by (a) voice and (b) digital, with their own "parallel" WINLINK system, without any of the normal interference of the amateur radio bands. The SHARES teams are divided up by Regions (we are in Region 4) and activate for any important disaster or event. FL BAPT DR and some others hold SHARES federal licenses (from the NTIA, not the FCC) You may be able to make connections via this very useful system. Their voice nets run 24 hours during disasters. I've reached them in the wee hours of the morning with a hurricane passing overhead.... Their WINLINK servers operate 24 hours a day, continuously.

JS8 CALL
This innovative system uses extreme weak-signal techniques but provides store-and-recall messaging, and relay capabilities. Interest in using it for emergency communications is growing. Software is free, and uses a soundcard. The modulation system is identical to that of FT8 and the software is similar, but more adapted for normal contacts and conversations. See: http://js8call.com/

LOCAL COMMUNICATIONS

A major goal for local communications is maintaining communications with volunteers doing ministry or security.

Repeaters: Amateur Radio Directories
If a local repeater in the area of ministry still works, and you have deployed units with amateur radio volunteers, the repeater may be an excellent resource to help keep in touch with them.

Many amateurs are familiar with a "repeater directory" that lists available VHF/UHF repeaters. It would be a very wise idea to print out the applicable pages from such a directory for the area of service BEFORE HEADING THERE. Some of those repeaters may still be in operation; others will not. Some users may congregate (using simplex) on the "output frequency" (user receive frequency) of the repeater and have useful information.

Bringing Your Own Repeater: You may be able to provide a temporary "replacement repeater" for the area, as your disaster ministry gains assets. Remember, the key for a repeater isn't really power -- it is HEIGHT.

Direct Simplex Radio
For local communications, for example to keep track of volunteers, you are likely to use amateur radio (if the participants are licensed) or voice business band frequencies for which Florida Baptist or a related organization has a purchased license and legal ability to use the frequency. Regular hand-held and mobile transceivers are the of choice for these tasks.[6] Even GMRS might come into play.[7] Remember that HEIGHT OF YOUR ANTENNA is the primary "range" determining factor, so getting a VHF/UHF antenna at the top of a mast, or up high with a tree using a slingshot will make a huge difference.

Here is an example of a topical conversation with a deployed unit over a business band frequency (material retyped with small changes from

---

6   On business band frequencies, these will be Part 90 certified radios that are programmed by a professional radio shop qualified to perform that work. Only in true dire situations should any other equipment be utilized on these frequencies.
7   If licensed to individuals or families.

Missouri Baptist Disaster Relief):

TYPICAL CONVERSATION

Com Trailer on 2 meter Amateur Radio: "W0GRP Mobile 2, this is KD0KVS Baptist Com."

Mobile 2: "KD0KVS Baptist Com, this is W0GRP Mobile 2, go ahead."

Com Trailer: "Operations needs to know the current location of Recovery 35."

Mobile 2: "Operations need to know the current location of Recovery 35, Stand by Baptist Com and we will check. W0GRP Mobile 2 clear."

Com Trailer: "That is correct, KD0KVS Baptist Com standing by."

Mobile 2 on Business Band: "Recovery 35, this is Mobile 2."'

Recovery 35: "Recovery 35, go ahead."

Mobile 2: "Operations need a current Work Order Number where your unit is located."

Recovery 35: "Operations needs a current Work Order Number where our unit is located. We are current at Work Order Number 135."

Mobile 2: "I understand. Recovery 35 is currently at Work Order Number 135." Thank you, Recovery 35. Mobile 2 clear."

Recovery 35: "Recovery 35 clear."

Mobile 2 on 2-meter Amateur Radio: "KD0KVS Baptist Com, this is W0GRP Mobile 2."

Com Trailer: "W0GRP Mobile 2, this is KD0KVS Baptist Com, go ahead."

Mobile 2: "Baptist Com, Recovery 35 is current at Work Order Number

135.:

Com Trailer: "Recovery 35 is current at Work Order Number 135. Thank you, Mobile 2. KD0KVS Baptist Com clear."

Mobile 2: "W0GRP Mobile 2 clear."

Notice what extreme effort was made to ensure accurate transmissions, with multiple redundancies employed. Also notice that no personally identifiable information (addresses, etc) were transmitted. Those are good techniques to use to protect our clients and volunteers.

UNIQUE MINISTRY OPPORTUNITY: PROVIDING SURVIVORS A WAY TO REACH LOVED ONES

The concept of (outgoing) Health and Welfare traffic from survivors reaching out to reassure worried loved ones "back home" has a long and rich tradition in the Amateur Radio history, particularly with the National Traffic System, which has several "ARL" texts that specifically address this type of communication.

In recent years, it seemingly has received less attention, yet the need still remains. What is necessary in order to provide this important and valued ministry[8] includes:

- Some direct contact with survivors allowing them to fill in a form to provide message details including addressing information, to send back to loved ones;

- Some method by which those messages can be efficiently forwarded.

We now have the capability to accomplish BOTH of these requirements as follows:

Capturing Messages
This can be done as simply as having blank paper forms that can be filled out by survivors at feeding stations operated by Florida Baptist Disaster Relief --- or by associated NGOs (such as the Red Cross) and then returned expeditiously to the disaster ministry. (Red Cross may have satellite systems up and running and able to utilize their "Safe and Well" system If so, then the system described here isn't needed; but there are a limited number of satellite systems, and you may well find that many volunteers aren't reached by those existing.)

---

8   James emphasizes the practice side of faith: "If a brother or sister is
     without clothing and in need of daily food, and one of you says to them,
     "Go in peace, be warmed and be filled," and yet you do not give them what
     is necessary for *their* body, what use is that? Even so faith, if it has no
     works, is dead, *being* by itself." James 2: 15-17

Forwarding Message

By far the easiest method to forward these messages is via WINLINK. **Capturing an email address for delivery makes it almost direct and immediate via the WINLINK system** (*your message will reach the email account of the intended recipient within seconds of completion of your transfer*) -- however, there are even possibilities to forward suitably formatted messages into the Red Cross Safe and Well system, which acts like a giant bulletin board system and can be searched by multiple family members. Either can be handled by the WINLINK System.

But there are even more systems: the NTS-D can accept short messages to family members, and standard RRI or NTS voice nets can also accept such messages.

An example of a "form" inside WINLINK that allows for easy creation of outgoing welfare messages. You do NOT have to use this -- a simple email message can easily be created with normal WINLINK.

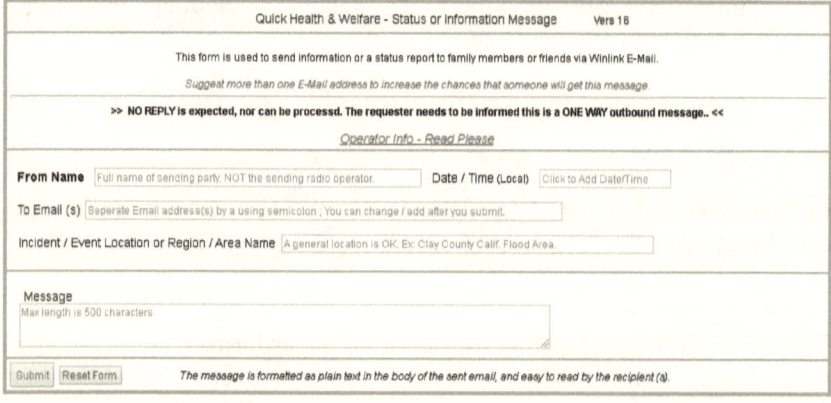

The next page is a complete example of a paper form that might be used to capture messages from Survivors and then entered into WINLINK to send back to family follows on the next page -- photocopy as needed:

See Appendix Three for an explanation of how to leverage multiple typists (who do not have to be amateur radio licensees) and computers to speed the data capture from these paper forms.

OUTGOING MESSAGE TO LOVED ONES
VIA _____ DISASTER RELIEF

MESSAGES MUST HAVE AT LEAST EMAIL ADDRESS -or- PHONE NUMBER FOR DELIVERY	
FROM:	PRINT YOUR FIRST NAME	LAST NAME
DATE:	MONTH DAY, YEAR
INCIDENT:	GENERAL NAME OF INCIDENT
TO:	PRINT PERSON ADDRESSED TO
	PRINT EMAIL ADDRESS
	REPEAT PRINT EMAIL ADDRESS
	PRINT **PHONE NUMBER** INCLUDING AREA CODE ( )
MESSAGE	PRINT MESSAGE TO SEND (MAX 100 WORDS) IN LINES BELOW:

MESSAGE CREATOR ACKNOWLEDGES THAT ACCURACY OR EVEN DELIVERY IS NOT GUARANTEED   THERE IS NO CHARGE FOR THIS SERVICE DONE BY VOLUNTEERS.
DO NOT PUT EMBARRASSING OR PRIVATE DETAILS AS THIS MESSAGE WILL NOT BE ABLE TO BE KEPT CONFIDENTIAL.

SATELLITES

SATELLITE DIGITAL COMMUNICATIONS -- COMMERCIAL

Satellite systems can offer amazingly high speed internet connections in a completely devastated area. (Performance may be 4 Mbit/second or faster, shared with a few other users to reduce costs.) However, they have their own unique difficulties, and more frequently than wished, they don't work.

First, the system for aiming must be working and accurate; these are tiny signals. A high-gain antenna, and low-noise receiver and modem[9] are necessary to decode and provide signals.

Satellite communications are quite different from terrestrial internet and have their own difficulties as a result.

BIT ERRORS
The bit error rate (BER) on a terrestrial (fiber, copper) system is typically below 10^{-10} but on a satellite link, the bit error rate can be many orders of magnitude greater, as much as 10^{-2} to 10^{-6} . This necessitates extensive Forward Error Correction, which reduces the channel capacity.

LATENCY
Latency is how long it takes to get some response. On terrestrial links the latency is in the range of 30 milliseconds (30 ms) and for links at close distances, may be much less. But the extremely long distances the radio waves have to travel up and back from a geostationary satellite cause round-trip time (RTT) latencies of about 520 milliseconds. This can wreck havoc with normal TCP/IP communications, resulting in unnecessary retries and excess packet transmission. Software to "adjust" the protocols utilized may be needed. These systems may increase the window size of packets, and use large buffers to accommodate.

---

9 Modem: just as with your cable system at home, the modem takes digital characters from the Ethernet local area network in the comms trailer and creates electrical signals that are appropriate for the satellite transceiver, and *vice versa*.

DATA RATE ASYMMETRY

The up link may be lower power than optimum, and therefore the ratio of the speeds of up link and down link may be great, as much as 100:1. This causes real problems with "acknowledgments," which are a key part of the lower layers of the ISO network layer model of communications.

Special software in the satellite data system may be utilized to attempt to mask some of these shortcomings. The output of the modem or higher layer system then becomes available within the communications trailer as a normal tcp/ip network connection.

This satellite data connection will typically be split out to multiple computer users, with a small **router**, often using DHCP to assign internet protocol numbers to each user computer. This works similarly to the router you probably have in your home to split out your cable modem to multiple users.

The amateur radio operator may be of considerable help in managing satellite systems after a bit of time spent "reading the manuals." It would be a good idea for volunteers who expect to work in the communications trailer to gain considerable familiarity with:

- tcp/ip protocols, ports, ip numbers
- commands such as IPCONFIG IFCONFIG
- testing commands such as PING, NET STAT, ARP, TRACERT (traceroute on linux systems)
- DHCP settings on a typical router

One can easily read about these areas at leisure at home, and even practice them on one's home network system.

4 VOICE COMMUNICATION TECHNIQUES

Voice communications may be over HF SSB or VHF/UHF FM. They may be simplex, or via a repeater, or as part of a "net." Voice communications are ideal for short, terse communications or discussions, often referred to as "tactical", e.g.,

"Mud1, this is Base, what is your current location?"

But they can also be used for formal written (generally short) traffic as well.

The American Radio Relay League EC-001 course makes the important point that written communications are very important when some significant resource or action is being requested. Justification, tracking, and administration are much more easily handled when you send a formal message asking for two thousand pounds of butter, instead of just a quick, "Jim, can you get us 2K of butter?" This was a lesson learned hard all over again in Hurricane Michael by many volunteers. *If it isn't written, it may never be carried out.*

While the ICS-213 is an important government document with which all radio volunteers should be familiar, it was originally developed for hand-carried notes, and does not have all the preamble tracking information of the more venerable ARRL Radiogram. It is a good skill to recognize that an ICS-213 can be disassembled easily and the relevant parts sent inside a simple ARRL Radiogram -- with an "op note" (operational note) to present it to the intended recipient "put back together again" as an ICS-213. (see written example later in this section)

Therefore, all disaster radio volunteers should be quite familiar with the simple radiogram format and able to both send and receive these efficiently over voice.

That turns out to be be more difficult than one might think, and after adequate training on voice procedures, many volunteers have a new appreciation for why DIGITAL (or even Morse code) is often a cleaner technique where appropriate.

SUGGESTION: Make photocopies of these forms rather than using them right in the book. After a little practice, you won't need the form at all -- you'll just use blank paper.

--

NR	PRECED	HX	Stn of Origin	Check	Place of Origin	Time Filed	Date Filed

Amateur Radio "Radiogram"

Addressed TO:

Message Received At:
Station: _____ Phone: _____
Name/Addr: _____

email_____

phone_____

OP NOTE:

<BT>

_____ _____ _____ _____ _____   _____ _____ _____ _____ _____
_____ _____ _____ _____ _____   _____ _____ _____ _____ _____
_____ _____ _____ _____ _____   _____ _____ _____ _____ _____
_____ _____ _____ _____ _____   _____ _____ _____ _____ _____

<BT>
SIGNATURE

OP NOTE:

RCVD FROM	DATE	TIME	SENT TO	DATE	TIME

Amateur Radio "Radiogram"

NR	PRECED	HX	Stn of Origin	Check	Place of Origin	Time Filed	Date Filed

Addressed TO:

Message Received At:
Station: _____ Phone: _____
Name/Addr: _____
_____
_____

email_____
phone_____

OP NOTE:

<BT>

<BT>
SIGNATURE

OP NOTE:

RCVD FROM	DATE	TIME	SENT TO	DATE	TIME

Key Points on Radiograms:

- Everyone **knows** what is in the preamble -- so just read them, (CORRECT EXAMPLE): "Please copy number 1, Routine, KX4Z, 12, Gainesville Florida Oct 12" -- no one needs to hear

- (INCORRECT EXAMPLE): "Please copy number 1, Precedence Routine, Originating Station......"

- PAUSE (say "Break"and let go of the mic button so you can hear the other fellow!) between the address and the text and between the text and the Signature -- let the other fellow ask for "fills" if they need.

- Use phonetics and procedural words (see below) to help out.

- GO SLOW!! The other fellow is trying to both HEAR and WRITE --- and no one can write very fast.

Important Procedural Words ("pro words")

I SPELL
FIGURES   (234)
INITIALS   (ABC)
MIXED GROUP (e.g. 1A2C)
AMATEUR CALL SIGN
WORD AFTER
WORD BEFORE
ALL BEFORE
ALL AFTER
STANDBY

--   becomes DASH
@   becomes ATSIGN

ITU Phonetic Alphabet

Letter	Word	Pronunciation
A	Alfa	"AL-FAH"
B	Bravo	"BRAH-VOH"
C	Charlie	"CHAR-LEE" or "SHAR-LEE"
D	Delta	"DELL-TAH"
E	Echo	"ECK-OH"
F	Foxtrot	"FOKS-TROT"
G	Golf	"GOLF"
H	Hotel	"HOH-TELL"
I	India	"IN-DEE-AH"
J	Juliet	"JEW-LEE-ETT"
K	Kilo	"KEE-LOH"
L	Lima	"LEE-MAH"
M	Mike	"MIKE"
N	November	"NO-VEM-BER"
O	Oscar	"OSS-CAH"
P	Papa	"PAH-PAH"
Q	Quebec	"KEH-BECK"
R	Romeo	"ROW-ME-OH"
S	Sierra	"SEE-AIR-RAH"
T	Tango	"TANG-GO"
U	Uniform	"YOU-NEE-FORM"
V	Victor	"VIK-TAH"
W	Whiskey	"WISS-KEY"
X	Xray	"ECKS-RAY"
Y	Yankee	"YANG-KEY"
Z	Zulu	"ZOO-LOO"

Number	Pronunciation
0	"ZEE-RO"
1	"WUN"
2	"TOO"
3	"TH-UH-REE" or "TREE"
4	"FOW-ER"
5	"FI-IV" or "FIFE"
6	"SIX"
7	"SEV-EN"
8	"ATE"
9	"NIN-ER"
Decimal	"DAY-SEE-MAL"

You are going to need to PRACTICE that phonetic list!!! Don't be lazy and make up your own. When you are driving down the freeway, voice out the road-signs using those phonetics -- and even an old CW op like myself can get used to them!

PRECEDENCE	Comment
EMERGENCY	spell out on CW and RTTY
PRIORITY	P on CW
WELFARE	W on CW
ROUTINE	R on CW

OPTIONAL HANDLING INSTRUCTIONS	
HXA (followed by number)	Collect landline delivery within X miles. If no number, unlimited authorized.

HXB (followed by number)	Cancel if not delivered within X hours of filing time; service message to originating station
HX C	Report time and date of delivery to originating station
HXD	Report to originating station the identify of station from which received, date and time. Report identity of station to whom relayed, plus date and time, or delivery date time and method.
HXE	Delivering station get reply from addressee, originate message back.
HXF(followed by number)	Hold delivery until date
HXG	Delivery by mail or landline toll call not required; may cancel message and service originating station.

Important Net Etiquette:

1. Follow the Net Control's Instructions
2. Don't disappear -- stay there until released

ICS-201 DOCUMENT

GENERAL MESSAGE		
TO:	POSITION:	
FROM:	POSITION:	
SUBJECT:	DATE:	TIME:
MESSAGE:		
SIGNATURE:	POSITION:	
REPLY:		
DATE: TIME:	SIGNATURE/POSITION:	

ICS-201 DOCUMENT

GENERAL MESSAGE		
TO:	POSITION:	
FROM:	POSITION:	
SUBJECT:	DATE:	TIME:
MESSAGE:		
SIGNATURE:	POSITION:	
REPLY:		
DATE: TIME:	SIGNATURE/POSITION:	

HOW TO SEND AN ICS-213 AS A RADIOGRAM
A VERY IMPORTANT SKILL

Amateur Radio "Radiogram"							
Fill in the preamble just as you normally would use the date and time from the ICS-213							
NR	PRECED	HX	Stn of Origin	Check	Place of Origin	Time Filed	Date Filed

Addressed TO:

Put the TO entity and their POSITION HERE

email_____

phone_____

Message Received At:

Station: _____ Phone: _____

Name/Addr: _____

_____

_____

OP NOTE: PLEASE DELIVER AS ICS-213

<BT>

PUT THE SUBJECT AND MESSAGE TEXT HERE

<BT>
SIGNATURE
PUT THE SENDING OFFICIAL AND THEIR POSITION HERE

OP NOTE:

RCVD FROM	DATE	TIME	SENT TO	DATE	TIME

That OP NOTE "PLEASE DELIVER AS ICS-213" TELLS THE DELIVERING STATION TO REFORMAT INTO ICS-213 FORMAT. EASY!

5 DIGITAL (WRITTEN DATA) COMMUNICATIONS TECHNIQUES

VOLUNTEER	Training Goals for Digital Amateur Radio
Incident Command Communications Trailer	Must have a solid understanding of WINLINK HF (and possibly VHF) -- <u>to the level that they can troubleshoot difficulties when in theater.</u>
All ham volunteers	Have a solid understanding of what the WINLINK system can, and cannot do -- basically that it is useful for modest-sized (<5kbyte) messages or files, but NOT for large PDF or large spreadsheet files for example.

This Chapter can only scratch the surface of the WINLINK system. Volunteers need to put in the time [10] to learn about many of the features of this best-in-class 2-decade old system used by many NGO's and even governments. The more you invest here, the better volunteer you will be!

IMPORTANT:

<u>Before a deployment</u>, someone needs to configure the Windows WINLINK computer for the approximate "My Grid Square" Maidenhead Locator of the deployment location. (It doesn't have to be exact -- just roughly the city/state) -- and then UPDATE THE

---

10 Do you see someone skilled in their work? They will serve before kings; they will not serve before officials of low rank. Proverbs 22:29

RMS LIST. SETTINGS | WINLINK EXPRESS SETUP | UPDATE

Google "maidenhead grid square" and you'll easily find many web pages that will look this up for you.   Learn how to do this! **A table of useful Florida cities is in Appendix Two -- just picking a nearby city will be good enough for the propagation table calculations (the primary usage of the location information).**

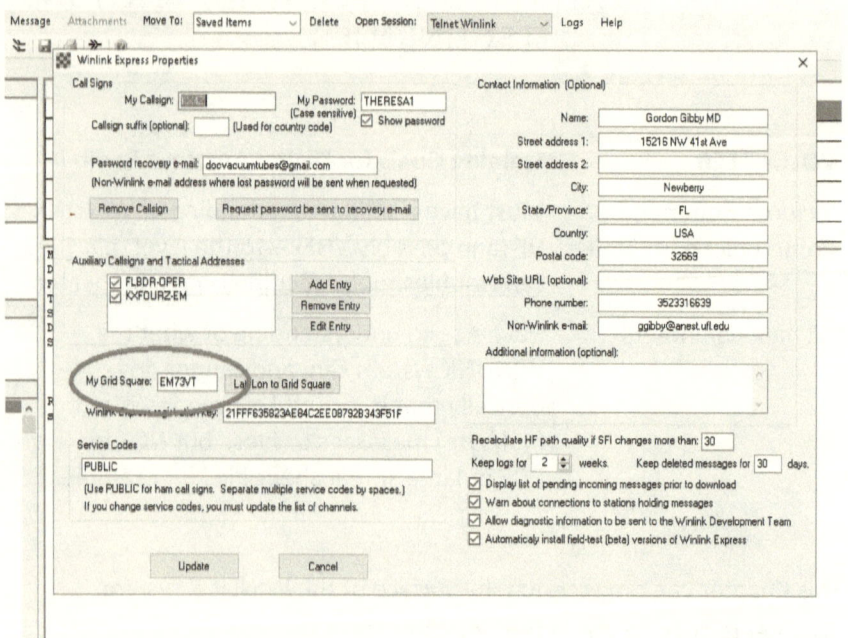

**Setting the Grid Square for the deployment Location
Important and far easier to do BEFORE you get there.**

IMPORTANT THINGS TO KNOW ABOUT USING ICOM 7300 SSB TRANSCEIVER FOR WINLINK

Receiver settings	Turn all noise blankers, compressors, narrow filters, and DSP *OFF* You want a very simple receiver. Let the PACTOR

	modem or other TNC Software do all the adjustments
CI-V Number (used to identify various ICOM radios for controlling the radio over their 1-wire bus)	7300: uses 94 hex (input 94) BAUD RATE: AUTOMATIC; select 9600 and it usually works fine.
Mode	USB - Digital on all bands!
Modulation	Expect the modulation to be delivered via the 13-pin BACK PANEL CONNECTOR. (It may be VERY wise to disconnect the microphone if exclusively doing WINLINK to avoid room noise being transmitted on a data frequency)
POWER	35-50 watts peak output is the sweet spot from a 100-watt transceiver. AVOID OVER-DRIVING. You do NOT want a bunch of "ALC" showing up.
OTHER DETAILED INFORMATION	**SEE APPENDIX ONE**

There are several YouTube videos that can also be helpful:
What is Winlink? by K4REF:
https://www.youtube.com/watch?v=qGhUfW8pjY8
Using a Sound modem TNC... by K4REF:
https://www.youtube.com/watch?v=RF0OMNZCEVA
Setting up sound levels.... by K4REF:
https://www.youtube.com/watch?v=7G1DBs-04MM

1. **COMPUTER:** The Comms Trailer has a fast Windows-based computer which has USB ports along its side(s). You will use one of those ports to plug in the ICOM 7300, and another to connect to the PACTOR modem. They will be marked, and it makes it much easier if you don't mix them up.

2. **USB CABLE:** The USB cables provided are shielded cables and

should have ferrite beads at both ends with a turn of cable in the beads to further suppress RFI. RFI interference occurs when "common mode" currents flow over such cables from one device to another, causing their "grounds" to literally be at different RF voltages, nanosecond by nanosecond -- and can cause software to crash. Ferrites and loops discourage common mode currents by inserting an impedance that only applies to common mode currents. (If the USB port ever does freeze, close the application, remove the USB cable from the computer port; reinstall the USB cable into the computer port and re-start the program.)

3. **RADIO CABLE**: The PACTOR modem must then be connected to the ICOM 7300 HF transceiver. A special cable with the 9-pin DIN connector for the PACTOR modem and the 13-pin connector for the rear of the 7300 is provided. BE NICE TO THOSE CONNECTORS

> Note: audio transmit and receive levels are set inside the WINLINK software

Setting	Value
FSK level	
PSK level	

4. **Interface Software: For HF operation, you only need the Winlink Express software. (Should you try VHF, you'll likely need additional software, not covered in this manual.)**

5. **WINLINK EXPRESS.** Start up the WINLINK EXPRESS software. I suggest that it be positioned in the left half of the computer screen.

6. **CALLSIGN:** After starting the WINLINKEXPRESS software, at the upper left corner you will need to select the proper CALLSIGN for your operation. If your callsign isn't listed, you can add yourself and run through the setup information which mainly involves putting in your WINLINK password and adding some location information about yourself. When your DISASTER RELIEF CLUB comes online and has its own callsign, operation will shift to this callsign for clarity. If you have a SHARES call sign and wish to use that privilege, select the SHARES call sign -- different channel choices will be made available to you. Ham radio callsigns are NOT allowable on the SHARES frequencies.

    NOTE: there is a special procedure for creating an account when you are not able to connect by way of the Internet to register your password. This will be explained in hands-on training.

7. **OPTIONAL REGISTRATION:** If the software requests a "registration" number, just wait a moment and it will give you the option to "remind me later"

8. **Writing an Email Message.** If you wish to create an email, that option is under **Message | New Message** and works pretty much like any email system. There is a contacts list that you can access by clicking on TO: or CC: buttons. Emails allow attachments, but the size may be limited to 40K (or some other number) due to the limited bandwidth available. TRY TO KEEP ATTACHMENTS DOWN BELOW 10 K!!! Once you are finished with your email, be certain to **Post to Outbox**, which queues it up to be transmitted on the next connection to a RMS (radio message server). You should see the number in the OutBox increment by 1. **<<-- LOOK FOR THAT!** It is great clue to watching what is happening!

WARNING: Be very judicious in sending any email to a regular internet ("smtp") email address By doing so, you give the person the ability to send YOU email <u>by way of amateur radio</u>. Obviously foul language, commercial activity and other non-appropriate content could be sent to you over amateur radio as a result---and there are plenty of people WATCHING for this on the very open, national WINLINK Viewer. <u>So limit this to only trusted officials or agencies for emergency usage, and remove any commercial vendors when the emergency is over. Study the WINLINK "white-list" to better understand this.</u>

The COMM UNIT LEADER should scrub the WHITE-LIST appropriately using internet access to your account at winlink.org after the disaster is over.

9. **Selecting PACTOR Winlink (or other modes):**   In order to connect over HF, you will select PACTOR WINLINK in the "Open Session" drop down, and then click the **CONNECT** icon (which looks like a pin connecting to a jack, or an arrow catching up to the tail fletching in front) or click on **OPEN SESSION.**

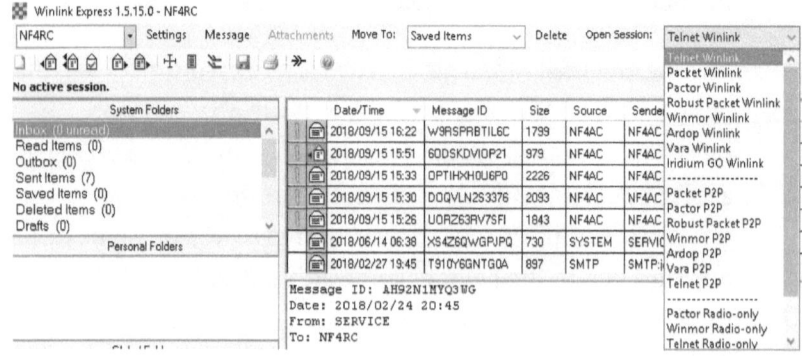

WINLINK has a LOT of ways to send messages.   In this short text we cannot hope to explain all of them; you're going to need hands-on training or a lot of reading.   The important ones for us to get messages out of a disaster area are typically:

- PACTOR WINLINK
- ARDOP WINLINK
- WINMOR WINLINK

For this text, we deal ONLY with communications to a RMS station out of the disaster area, which can put your message directly on the Internet, and retrieve messages for you, from the Internet. We don't deal with "peer to peer" or "Radio-Only" -- more advanced and great techniques for specialized issues.

10. **Settings:** **We should have the correct settings for the Pactor Modem and the Radio already entered. If you had to use a sound card technique (e.g., ARDOP or WINMOR) you might have to select the 7300 radio/codec in case the computer "forgot" the radio.**

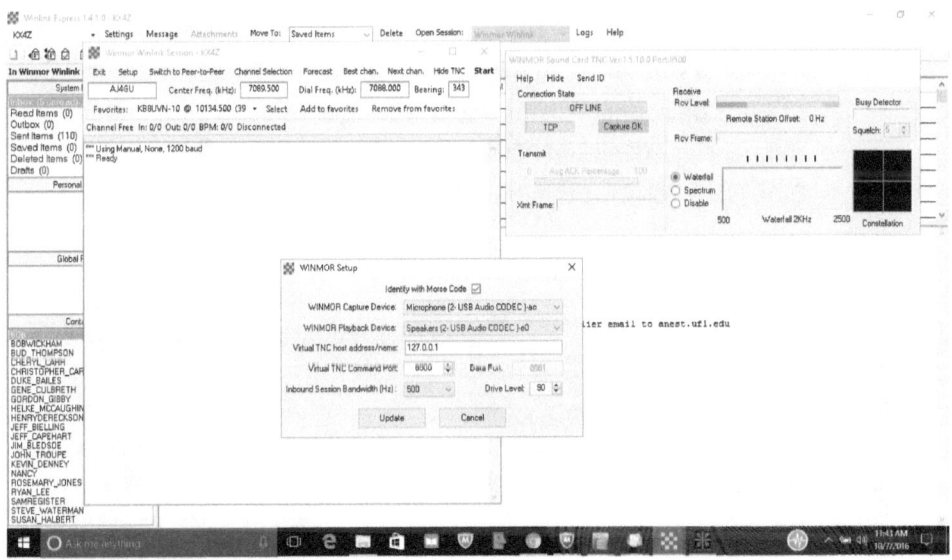

11. **Choosing an RMS to try to contact:** Once you have the Session dialog open, click CHANNEL SELECTION to get to the propagation prediction screen (which knows all the available winlink servers and displays which are easiest to reach at the

current time, and on which bands):

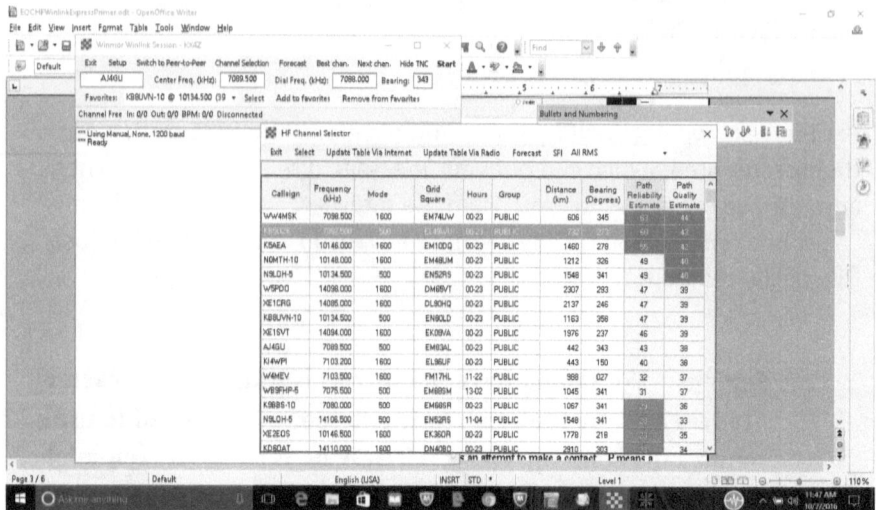

This works much like a spreadsheet -- you can sort it by Path Quality or frequency or alphabetically by station --- become familiar with this!!!

12. **Click on a station you'd like to connect to.** You'll be returned the session connect dialog and it will show you two different frequencies: CENTER FREQUENCY and DIAL FREQUENCY. The DIAL FREQUENCY is the important one for you!

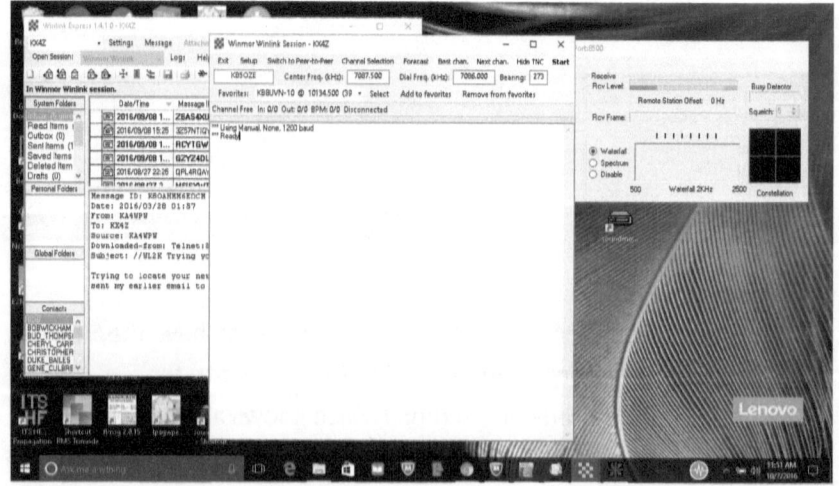

13. If the radio is correctly configured, it will snap to the proper frequency. If not, you're going to have to dial it to the DIAL FREQUENCY. Hint: the "center frequency" is always equal to 1.5 kHz ABOVE the "dial frequency" -- and remember, that all WINLINK is in UPPER side band, even on 80 and 40 meters.

14. **TUNING ANTENNA:** We will have an auto-tuner. It requires a modest amount of RF power to tune. It may be automatically chosen by pressing a TUNE option --- or by choosing CW and pressing a Morse code key. Learn how to listen for the relays clatter and recognize when the SWR is properly adjusted. (Consider reading the manual.)

15. **Making a connection: START** When you Click **START**, you should see the system begin to transmit and send digital tones in an attempt to connect to the RMS. The output power on the ICOM 7300 should not hit into the RED region but should be near the upper portion of the "good" range.

16. **Try Try Again.** HF is not like VHF. Making a connection is NOT a guaranteed thing. Currently the antenna at the EOC is (to put it politely) "less than optimum". I am generally UNABLE to contact most winlink stations on their equipment that I can easily contact at home. Hopefully this will change. However, even with a good antenna, you may need to try a couple of stations/bands until you hit upon one with good ionospheric conditions for a contact. Go back to CHANNEL SELECTION (11) above and give it another try. You'll get very quick at this after a little practice.

EXPERIENCE AT HF IONOSPHERIC PROPAGATION IS KEY HERE.

Your first day on deployment, every two hours you should be testing for the strongest RMS stations you can reach and filling out a table for use should it become necessary. Here is an example:

TIME (local)	BEST STATIONS	TIME	BEST STATIONS
0000		1200	
0200		1400	
0400		1600	
0600		1800	
0800		2000	
1000		2200	

YOU NEED TO DO THIS EVEN IF OTHER SYSTEMS SUCH AS SATELLITE OR
CELL PHONE ARE WORKING.   THEY MIGHT QUIT!
This form is repeated in Appendix Three.

Map to find other packet stations:

http://www.winlink.org/RMSChannels   (Click the WINMOR option.)

(In order to get the SHARES list you will need to enter a special service
code, not printed here for security reasons.)

NOTE: On VHF, you are very unlikely to reach a packet station farther
than 6 or so miles unless your antenna is VERY VERY high.   The typical
"reach" of your antenna (in miles) is 20% greater than the square root
off the antenna height in feet. This is due to the curvature of the earth
and no antenna can get around it.   Read of the "other side" antenna
adds to this.

WHAT HAPPENS TO YOUR EMAIL:

You would be using this software for real in the circumstance that no
Internet is available in your area due to any of several causes. Once you
make contact and send your email to a distant RMS over HF, it will then
be immediately put on the Internet at that (unaffected) station or
alternatively, put up for further HF forwarding. It is a good move to

read the logon script information from the station you're connecting to as their script may tell you whether they have internet or are operating in "RADIO ONLY MODE" If the latter --- you might be better to switch to a station that still has internet connectivity, or your message might "sit" for quite a while.

AREAS LEFT FOR THE READER TO LATER EXPLORE

1. Peer-to-Peer movement of traffic
2. How to obtain a new account in the field for a volunteer without internet access
3. Sending and receiving Forms
4. Specifics of adjusting sound levels to maintain linearity
5. The "Radio Only" system for dealing with complete Internet loss
6. The White-list (in general, ONLY allow trusted individuals to email you from outside the WINLINK system)

6 FLORIDA BAPTIST SPECIFIC COMMUNICATIONS ANTENNAS

SAFETY

Special Training will be required before volunteers can safely handle the 60 foot trailer-mounted tower. The State of Florida occasionally offers training on nearly identical AlumaTowers utilized by the MARC Units. **This training is 2-3 days long and is highly recommended. (IO-MARC)**

If significantly extended, the tower MUST BE GUYED. The Trailer must ALWAYS be stabilized appropriately. Current MARC units use extremely heavy duty stabilizing systems and have CONSIDERABLE weight on the trailer due to a cab, repeaters, generator et.[11] The GUYS on the MARC UNIT trailers are STEEL GUY WIRE. The FL BAPT tower is not as high (60 vs 100 feet) and likely not as HEAVY but still, extreme caution should always be utilized when working around the tower, hard hats, gloves, safety monitors, etc.

EXTREME CARE should be taken around this tower. Avoid potentially limb-threatening CRUSH INJURIES. **The tower should always be viewed as POTENTIALLY FALLING and if possible, persons should always stay as much out of the range of the tower, should it come down in any direction.**

HF ANTENNAS

FL BAPT DR has a commercially manufactured folded, terminated dipole antenna that presents a reasonable SWR on all bands --- but may be feeding a good bit of the power into the terminating resistive load on many frequencies. The published losses for antennas of this general type can be as much as 10 dB.

End-Fed Multi-band Resonant Antennas
FL BAPT DR also has multiple versions of end-fed, resonant half-wave (or harmonic) antennas that work against a ground rod. These antennas have multiple advantages for rapid-deployment disaster communications:

Relatively low loss -- 49:1 Balun has measured loss of < 2dB across a wide range of frequencies. Far lower loss than the non-resonant 9:1 designs.

- Antenna is usable on all multiples (both odd and even) of the fundamental frequency (where the wire is a half-wavelength).

- Antenna wire can be oriented in almost any way needed -- sloping to a high mount; up steeply and then horizontal ("L");

---

11 AlumaTower Trailer-Mounted Towers:
https://www.alumatower.com/trailer-tower-with-shelter/

near vertical, or inverted Vee (with a high support at the center of the wire, and both ends a few feet above ground)

- 135-foot end-fed wire allows operation on 3.5-4 MHz, 7 MHz, 10.5 MHz, 14 MHz, and most HF amateur bands above that.

- 100-105 foot end-fed wire allows operation on various SHARES HF frequencies, including those near 4.5 MHz, 9 MHz, 14 MHz.

- These antennas, fed from the end, present very high impedances to the Balun (2500-5000 ohms). The off-resonance capacitive or inductive reactance turns out to be a lower percent of the resistive radiation impedance -- giving these antennas a wider "swr-bandwidth" than one is used to, with a center fed, 50-ohm dipole.

Suggestions: Use a relatively short ground wire (no more than 3 feet) from the coax/ 49:1 Balun to a suitable ground rod to have effective performance. The ground is less critical because the impedance of the output of the 1:49 Balun is in the thousands of ohms. A tuner may still be required, but these types of antennas have losses around 1-2 dB (in the Balun) **and work on all harmonics of the original fundamental half-wave length**. The Balun toroid core can heat up with higher power --- avoid having that matching Balun in direct sunlight Even shielding it with an opened book or pamphlet will reduce its internal temperature considerably. For further information on this type antenna, consult:

Ad-Hoc Center Fed Dipole
In an emergency, a half-wave dipole can always be made, fed with 50 ohm coax at the center. The total length for the fundamental frequency, in feet, is equal to 468/frequency (MHz). A center fed antenna generally presents a good match to 50 ohm coax on the fundamental and **ODD** harmonics.

FL Bapt DR owns an auto-tuner, which can tune a random-length center fed (or fed almost any way imaginable) dipole, connected by either 300- or 450-ohm window line, which is very, very low loss. Avoid running such balanced window line right next to, or clamped between, metal -- that will increase the losses.

ANTENNA GEOMETRY
For HORIZONTAL antennas, whether center or end fed (makes no

difference) if the antenna is <u>below about 1/4 wavelength in height</u>, much of the power will radiate at relatively HIGH angles (> 45 degrees), favoring first-hop stations at relatively shorter distances.

For HORIZONTAL antennas, whether center or end fed (makes no difference) if the antenna is <u>ABOVE about 1/2 wavelength in height</u>, much of the power will radiate at relatively lower angles (below 45 degrees) favoring first-hop stations at relatively longer distances.
Horizontal antennas placed very closely to the "ground" (Florida soil is poorly conductive unless wet).....such as 5 feet or so....make excellent earth-worm warmers, and poorer sky wave radiators.

The azimuthal locations of horizontal antenna's major lobes is quite complicated and beyond the scope of this handbook. Suffice it to say, that if your antenna is less than 1 wavelength above the earth, you can expect to generally reach out in most directions.

VERTICAL ANTENNAS tend to be more omni-directional and sent more of their power out at VERY LOW elevation angles and are more useful for working HF DX. They are favored however for VHF where one wishes to reach nearby vehicles (which are nearly horizontally displaced from your location) and which almost ALWAYS use vertical antennas.

The TOWER owned by Florida Baptist DR can be used to support the center or the end of one of these HF Antennas. Note the moment arm provided by the horizontal pole at the top of the long mast at the top of the tower --- If you pull hard on that pulley, you are going to bend that mast! Be careful!!

VHF/UHF ANTENNAS

For the most part, VHF/UHF communications are direct line-of-sight, rather than refracted from the ionosphere. Because of the curvature of the earth, and the fact that radio waves don't penetrate DIRT at all, and houses and structures poorly, what limits communications distances the most is the HEIGHT of the two antennas involved. The maximum VHF/UHF range of a single antenna at X feet above ground, measured in miles, is approximately 20% greater than the square root of X. Therefore, a 60-foot high VHF/UHF antenna has a "range" of about 9 or

10 miles.

Coaxial feedline is the most common at VHF/UHF frequencies, but the losses are considerably higher than at HF. Rules of thumb:

Even for RG8X, at VHF, getting more height generally outweighs the increased losses in the coax up to 60 feet.

For UHF, one is much better advised to use thicker coax, such as LMR-400 size cable or RG8 rather than RG8X, and RG-58 should be avoided except for short patch cords.

Vertical antennas are the general rule on VHF/UHF basically because the majority of intended users (mobile or handheld) will have vertical polarization to their antennas. A simple dipole works well; a collinear dipole (two or more elements vertically aligned to flatten the elevation radiation profile more toward the horizon) can provide single digits of dB gain, which may make up for feedline losses.

TOWER MOUNTED VHF/UHF ANTENNAS
Typically VHF/UHF antennas will be supported with some form of "standoff" either from a vertical mast on the top of the tower, or from the top portion of the highest section -- because you cannot connect anything to the other sections that telescope inward. A 1 or 2 foot separation should be adequate, but remember there will be some "shadowing " of the signal due to the presence of the tower. Coaxial feed cable can either be carefully snaked all the through the center of the telescoping tower sections, or run on the outside of the tower. Be careful of wind loading from that coax if you allow it to be on the outside -- it could create some **significant moment arm on the mount or the tower.** Best to keep it fairly tight in to the tower. You want to strain relieve it by connecting it to the tower near the mount to avoid pulling your mount off the tower!

When attaching standoffs to the aluminum tower legs, with U-bolts, be very careful not to damage the aluminum tower tubes. Stainless steel hose clamps offer an alternative, but again don't overtighten. Manufacturers tend to build commercial standoffs out of very heavy material and charge a princely sum for them, but if you don't need the item to work during the peak of the hurricane, a much simpler solution can be used. Here is an example of how simple aluminum angle-iron can be used to mount two antennas on a tower:

Photo from KA1RCI.net -- a Silent Key from cancer, 2017.   But his work lives on, on the web.   Note how simply this can be fabricated.

A homemade mount with lots of options for VHF/UHF antennas has been fabricated for FL BAPT DR.

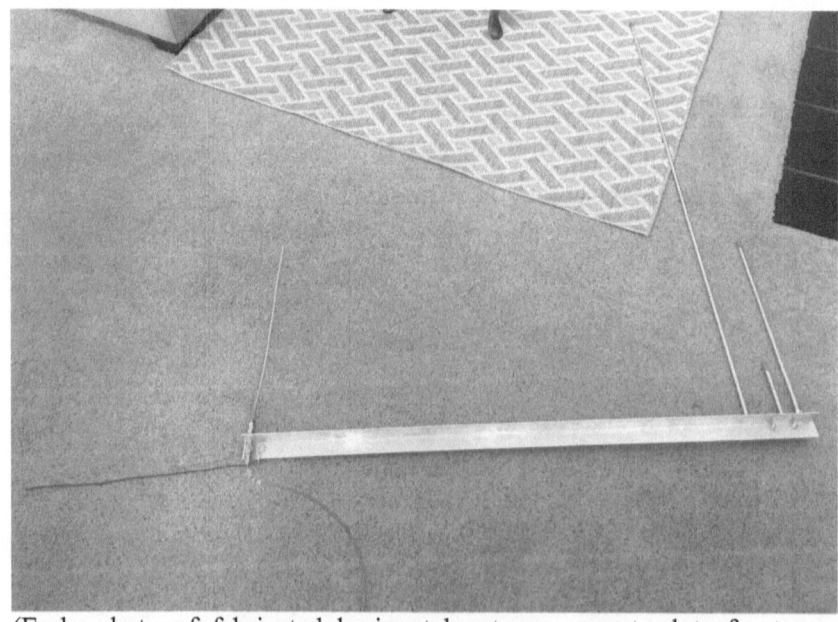

(Early photo of fabricated horizontal antenna mount, slots for tower connections added later.)

This type mount allows for multiple possibilities of antenna elements to be chosen:

- Dual VHF/UHF Arrow-band open-coupled J-pole -- heavy duty, causing a bit of moment to the mount. Can be mounted extending upward, or downward. If downward, a non-conductive support strut (or even string) can be connected from the center of the long element to the nearest tower leg. Only connect such a strut to the CENTER of that element, which is the low-impedance point at the fundamental frequency.

- Copper VHF element, and associated single radial -- two versions, one for 2 meters, and the other for 151-156 business-band frequencies. The "radial" provides the best SWR when extending horizontally in line with the mount.

- Shorter threaded rod UHF element, approximately 6" long for UHF ham or UHF business band. A tightening nut can be used if necessary to set the exact length for best SWR.

IF A TOWER ISN'T AVAILABLE

VHF/UHF antennas can also be pulled up to a good height using a tree, if no tower is available. Use rope or strong zip-ties or other non-conductive material to connect to their top end and simply pull the up by means of a rope over a high branch. Slingshot or other technique can be utilized to establish the line in the tree.

Gordon L. Gibby MD KX4Z

7 EMERGENCY HF ANTENNAS

I recommend your choice of any of THREE basic types of HF antennas that can be easily constructed and deployed. ALL of these antennas can be positioned horizontally, inverted vee, sloping, or vertical. A wise person once said, "put the most wire you can, as high as you can, in the clear of surrounding obstructions" and that is pretty good advice.

ANTENNA	Advantages	Disadvantages
Resonant half-wave dipole fed by 50 or 75 ohm coaxial cable Total length = 468/freq in MHz (in feed). Cut in half, solder to each side. Each side will be 234/F in MHz.	Easy to make, works on odd harmonics of design frequency	Will not work on EVEN harmonics. Formulas are great -- but you have CHECK the swr and often make adjustments. Bandwidth on 80 meters may be about 200 kHz
End fed half-wave horizontal, or sloping, or vertical antenna, using 1:49 Balun and a ground rod right there. Feed coax to the Balun.	Works on both even and odd harmonics. Ground rod does not have to be more than few feet because of the high impedance	Small loss in the toroid, about 1.5 dB SWR typically 1.5-3, usable, but may benefit from a tuner.
Center fed non resonant antenna fed with 300 ohm window	Generally will work on ANY frequency.	Modest lost in the tuner, perhaps 1 dB.

line and an automated tuner at the radio end. Typical total lengths of the antenna are 90-200 feet. A 1:1 or 1:4 Balun may optionally be tried between antenna tuner and the window line, right at the output of the tuner.		Requires understanding of how the tuner works. Sometimes requires a bit of trial and error to get the impedances to work out.

I **always** recommend a 1:1 un-un (a Balun designed to get rid of common mode currents) in the coax line between the tuner (if used) or antenna and the transceiver.   These can be purchased for < $40 or made for <$10.   It merely involves looping the coax or other wiring about 10 times around a #43 toroid core of suitable size.

The most important thing you can have for your antenna is either a simple SWR meter that you understand how to operate, or an antenna analyzer that you also understand how to operate.   I would always suggest that you have a 1:1 and 1:4 Balun in your kit, and a few ferrite clamp on beads.

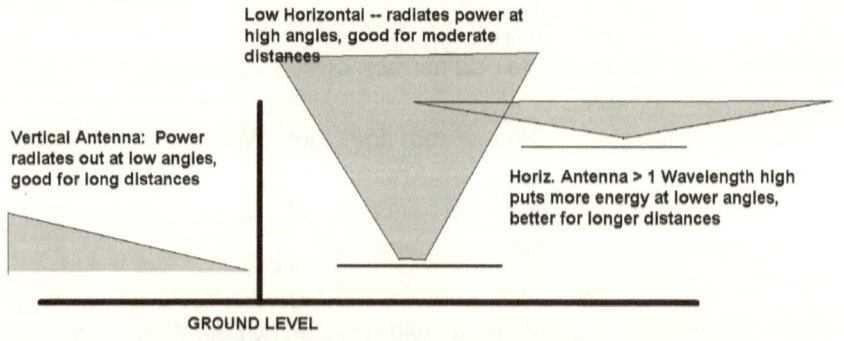

8 EMERGENCY VHF/UHF ANTENNAS

This is one of the simpler tasks in ham radio, so don't let it frighten you!!! Most mobile operators using VERTICAL antennas on their cars, so the traditional polarization for FM vhf (or uhf) voice work is VERTICAL.

A simple dipole is just two wires connected to the antenna end of some coax. If you use bare wire, the total length from one tip to the other) will end up being near

Length in feet = 468 divided by (Frequency in Megahertz)

so for the center of the 2 meter band

Length in feet = 468 / 146 = 3.2 feet = approximately 38 inches

(each side will be about 19 inches)

Just strip the coax to get to the center and shield conductors, connect them to two bare wires (you can twist-tie, crimp, solder, it doesn't matter!) slap on some duct tape to keep them kinda held in place, and suspend your antenna with a string from the top end. (It doesn't matter whether the center conductor goes to the top or bottom ends.) The COAX needs to go out at RIGHT ANGLES so it is uniformly exposed to both the top and bottom fields from the antenna to minimize unbalanced currents, ideally for maybe 10 feet or so --- and then you can make a simple homebrew "common mode inductance" (aka "Balun") by adding in about 4 turns of the coax in a 4" coil, and add some tape or zip ties or string to keep it in the coil.

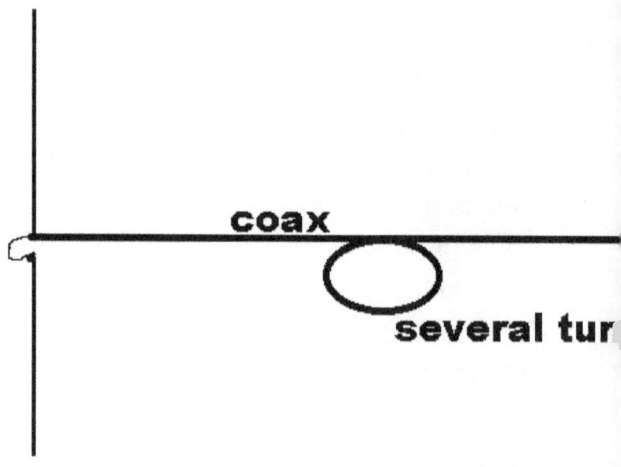

coax

several tur

Drawing: G. Gibby

Couldn't be much easier!

If you use INSULATED WIRE, or if you affix your bare wire antenna to a yard stick or something for support, it will require a slightly shorter length by a few inches. Taking an SWR meter with you will help for fine tuning but it isn't absolutely necessary as modern rigs are pretty well protected.

If you happen to have some FERRITES you can slip them onto the cable as well to reduce unbalanced currents --- none of that matters much if you are just doing voice work, but if there is a computer involved here....unbalanced currents can play havoc with Signalinks etc.

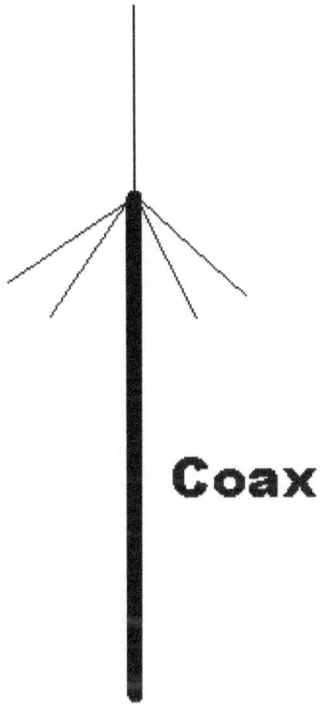

Drawing; G. Gibby

If you need to make an antenna fed with a VERTICAL coax run, then you can make a 1/4 wave vertical antenna with " 4 drooping Radials".   Run a vertical of about 19" (for the 2 meter band) upwards, and the add four evenly spaced 19" "radials"   at about 45 degrees from the vertical, and you'll reduce the induced unbalanced currents in the coax feed --- but you'll still probably wish to add the coil-balun or some ferrites to the line.

Pretty simple!!

9 BASIC HF PROPAGATION

Back in the "day" when most amateur radio operators communicated primarily on the HF bands, *everyone* understood HF ionospheric propagation. Now that many volunteers live in apartments and may have less experience with HF, it is important to go over a very few simple concepts.

IN GENERAL IN FLORIDA
HERE IS WHAT YOU ARE GOING TO FIND:

TIME	BEST WAY TO REACH Some One
Nighttime	80 meters for anyone within 500 miles 40 meters for persons at greater distances
Mid Afternoon	30 meters or 20 meters may be the ONLY usable bands for 2019-2023 Very close-in stations (up to 60 miles) with significant power may be reachable on 80 meters by near vertical incidence sky waves overcoming the significant absorption of the D-layer
Sunrise and Sunset (Gray Line)	Miraculous things may occur

Competing Effects
1. Florida poor conductivity soil means that "ground wave" (which works way better on 160 and 80 meters than higher bands, and WAY better with truly vertical antennas) *is not useful for more than a few miles.* If you were over sea water....it would be different.

2. We are near a sunspot minimum at this writing and that means the critical frequency (the highest frequency aimed vertically upwards that can sill be refracted a full 180 degrees and make it back down to someone within a very few miles of you) is down around 3 MHz at night, and may not exceed 5 MHz during the day.

3. Waves that approach the ionosphere at more oblique (more gentle) angles have a <u>better chance of being refracted back to hit earth</u>, and thus the geometry of farther-away-stations may work much more successfully -- hence creating a "skip zone" of persons in an annulus around you, extending from about 10 miles to about 100 or so miles....where it can be difficult to reach someone unless you use relatively long wavelengths.

4. <u>Waves of higher frequency are harder to refract</u> enough to come back to the earth; so for any given geometry, there is a MAXIMUM USABLE FREQUENCY for that particular geometry. (And the "skip zone" is bigger and bigger for higher frequencies. Good luck getting to ANYONE in your same state on 20 meters, for example.)

Practical Upper Frequency Limit

It is very important to understand the concept of SKIP ZONES and CRITICAL FREQUENCIES. These are two ways of explaining why NEAR stations cannot hear you, and explains the **UPPER frequency limit** for a local net. Trying to hold a small-state-net on a frequency *above the critical frequency* is a disaster -- no one hears anything-- unless they are several states away. *And the higher in frequency you go, the larger the circle will get, of people who cannot hear you. Plan on relays!*

---

5. The D-Layer (so-called "Destroyer layer") is heavily energized during the sunlight hours by the sun and can wipe out pretty much any wave below 40 meters and will try hard to wipe out 40 meter waves. 20 meters gets through relatively unscathed.

D-Layer Sets the Lower Frequency Limit

It is very important to understand the D-Layer ("Destroyer" Layer). Destroys **lower frequency** communications much more than higher ones -- driven by sunlight. This is the reason that state ham radio nets use 80 meters BEFORE the sun gets high and in the EVENING -- When the sun is high, 80 meters is wiped out *unless you are using weak-signal digital techniques.*

---

6. All of that added together -- and particularly the D-layer during the daytime -- **means that you will hear the strongest signal from a station at any given distance, when you are at a frequency JUST BELOW the maximum frequency that can still bounce back to earth from the ionosphere to that distance.** Thus

- time of day
- geometry of angles to the ionosphere
- frequency

all interact in important ways, and make it relatively important to understand how the ionosphere works both from theory and practice.

And it also means there may very well be NO FREQUENCY at HF where stations of every distance in your state can hear you at some hours....which is why relays have always been a part of HF ham radio, and why JS8's capabilities may become advantageous.

VHF AND UHF ARE VERY DIFFERENT: Other than rare sporadic E skip (Es) or tropospheric ducting, most VHF and UHF communications is simple LINE OF SIGHT direct radio wave travel. Because of the curvature of the earth.....HEIGHT OF YOUR ANTENNA is by far the determining factor. The distance at which your radio wave begins the impossible task of plowing through the dirt....is about 20% greater, in miles, than the square root of the height of your antenna in feet. So a 50 foot high VHF antenna begins to plow through the dirt around 9 miles out.

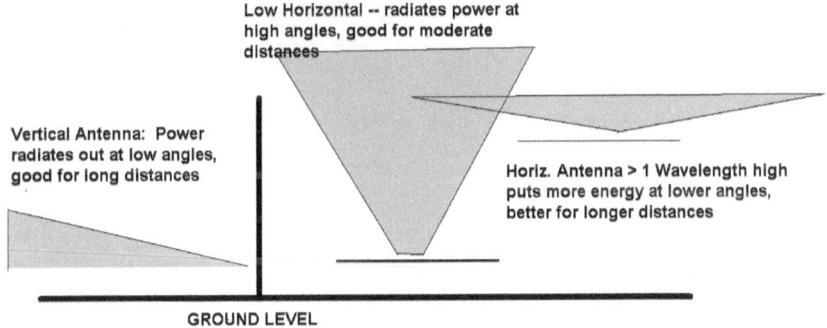

Stylized view of the elevation patterns of different types of antennas.

NOTES

10 ELECTRICAL POWER

In a disaster situation you are likely going to have to provide your own electrical power.   The Baptist DR folks may provide that with diesel generators, which will make your job easy.   Other alternatives include vehicle 12 volt batteries, and solar power systems.   Solar power systems are beyond the scope of this text because the Baptist DR will likely not need them.

One of the most important things for you to understand about emergency electrical power is **how to avoid catastrophic HF RF interference.**

Some forms of power generation can radiate so much RF hash on the HF bands that they make effective communications literally impossible.   We had to deal with that very issue at our Alachua County EOC, and I also have had to deal with it with my travel-trailer RV: inverter generators are generally a disaster, providing enormous radiated AND conducted wide-spread RF energy.   A mass of computers may be able to do the same.

How to deal with this?

FIRST: know your equipment.   You need to KNOW what the typical galactic "background noise'" is on the important bands, for normal antennas, on your own equipment, band by band, on the S-meter.   S-meters are horribly uncalibrated, so you need to know it for YOUR transceiver.   (The alternative is to know calibrated measurements on a spectrum analyzer, which we have done in Alachua County, but that is beyond most people.)

TYPICAL S-METER BACKGROUND READINGS ON FL BAPTIST DISASTER RELIEF ICOM 7300 CONNECTED TO A SUITABLE ANTENNA ON UPPER SIDEBAND

BAND	BACKGROUND S-METER
80 METERS	
40 METERS	
30 METERS	
20 METERS	

SECOND: If you identify that on deployment you are experiencing higher than normal amounts of noise, you will need to start hunting for and mitigating the source.

- If you find an offending inverter generator, you may have to arrange for it to be shut down during critical communications periods.

- If you find a vehicle with significant spark-plug-ignition noise, you may have to have it moved away from your antennas.

- You may wish to have "line filters" on your AC lines going to your radios to try and reduce conducted RF hash.

- You may wish to position your antennas AWAY from power lines or generators that may generate stick type noises.

- If your power source is not reliable, it can be a very good idea to have a high quality uninterruptible power supply that will kick in and keep the voltages going to your equipment stable --- but these are notorious RF hash generators also. The simplest solution may be to simply operate your equipment directly off a large 12 volt storage battery --- and then charge that battery during "off" periods when you don't care about any noise producers.

POWER POLE POLARIZED WIRING

I recommend that all 12 volt power systems be equipped with power-pole connectors, because these reduce the chance of a backwards polarity connection --- which can completely destroy an important piece of communications gear.

DIODE PROTECTION

I further recommend polarity protection systems, the simplest of which is a hefty silicon diode connected from the negative to the positive 12 volt wire, on the RIG side of the FUSE -- and connected so that it will ONLY conduct if the voltage is BACKWARDS -- in which case it will blow that fuse quick and protect the expensive radio from damage, by refusing to allow more than 0.8 volts of wrong-polarity power.   These are EASY to add to your power cables and more than one amateur wishes they had done so....

NOTES

11 STAYING LEGAL -- RADIO SERVICES

There are a dizzying array of different FCC radio services and it's important to choose properly to avoid accidental improper operation. Let's review their characteristics:

Radio Service FCC Regulations	Proper Users Licensing Equipment Power Antennas	Potential Service to Disaster Ministries
Amateur Radio Part 97	**Individuals.**-- not businesses **Licensed** after passing exam. Allowed to use vast array of equipment on proper frequencies, *even to program and alter equipment; repeaters allowed* Power can be beyond 1 kilowatt Main limitations on antennas relate to aircraft safety.	Cannot be primary technique for day to day business usage. Wide latitude to serve needs in disaster or emergency. With licensed individuals, serve both short- and long-range voice and digital communications.
Multi-Use Radio Service Part 95 Subpart J	Allows both **individuals and businesses.** "Licensed by Rule" -- means you don't have to get a specific license. Equipment must be pre-made; available in the $80 class for walkie talkies, for this service, non-adjustable -- 5 channels. 2 Watts maximum	Excellent choice for base communications in < 1 mile area. Possible for longer communications, must use tower antenna.

	Can use external antenna, up to 20 feet above structure or 60 feet above ground (whichever is greater)	
FRS Family Radio Service Part 97 Subpart B	**Individuals "for personal or for business use if you are not a representative of a foreign government " -- so OK to use for Disaster Relief groups** "Licensed by Rule" -- so no specific license required. Consumer un-modified UHF radios 1/2 - 2 watts, specific channels, usually handheld. 1/2 or 2 watts (recent change) Cannot add external antenna	**Excellent and very inexpensive choice for local base (short-range) operation**
GMRS General Mobile Radio Service Part 95 Subpart E	Currently intended for INDIVIDUALS (think: specific disaster volunteers) Must pay for 10-year license; usable by you and family members. Pre-programmed radios (channels) Power can be **as much as 50 watts** (note limitations on various conditions) **Antennas can be external and limited by aircraft safety**	**Can be extremely useful but only if individuals acquire the required licenses.**
"Business Band" Part 90 Private Land Mobile Radio Services (Industrial / business Radio Pool -- subpart C)	**Business licenses -- allow employees or associates to use radios.** Licensed by application. **Pre-configured handheld, mobile or base station equipment specific for this service.** (Not intended for non-professional adjustment.) Power -- typically up to 35	The most common type of radio utilized by businesses / disaster groups. Can be somewhat pricey ($250-$350) -- provide ongoing work for "radio shops" who professionally set up

watts output	systems for businesses.
Antennas -- wide array of limits, but generally **external antennas of considerable height are allowed**	Some inexpensive foreign handheld choices are available:
	https:// www.amazon.com/ Midland-Consumer-BR200-Business-Portable/dp/ B071Y8LN4M
	For higher power radios, see appropriate dealer online or in person.

The key take-away for amateur radio volunteers is to be very cognizant of the regulations for any given radio service. While amateur radio licensees are familiar with programming and using a vast array of gear in Part 97 operation, such consumer-modification isn't generally legal in ANY other radio service. Don't take on a task that legally must be performed by persons with an appropriate commercial license.

NOTES

APPENDIX ONE
ICOM 7300 WINLINK SETTINGS

The ICOM 7300 is an exciting new transceiver that brings fully digital transceivers to the general amateur radio market. Packed with capabilities, getting it configured for disaster ministry communications, including both voice and data can take a bit of work. This Appendix discusses the settings that seemed to work well for the ICOM 7300 in the Florida Baptist Disaster Relief Comms Trailer.

Best to understand the different ways the ICOM 7300 can accept signals to be transmitted:

Input	Signal Type / Examples
Front panel microphone connector	Voice over the microphone -- electret mic element generates low-level (millivolt) analog audio frequency electrical signals from speech. (If necessary, analog signals from a TNC or external sound card could be injected here)
Rear Panel accessory connector	"Line Level" (100 mV) analog audio frequency electrical signals frm a TNC or external sound card can be injected here -- and here is where we inject the signal from the Pactor Modem
USB (Universal Serial Bus)	The USB connection on the ICOM 7300 accepts digital data (1's and 0's) to give the 7300's **internal soundcard** the information needed to synthesize the proper audio signals for the digital mode desired. (The USB connection also allow setting the band and frequency and other settings.) This is how WINMOR and ARDOP are utilized by WINLINK -- and also how one can make PSK31, FT8, or any other "soundcard" type signal.

Just to be clear, this table shows how each type signal for transmission is connected:

TECHNIQUE	PHYSICAL INPUT
Single Sideband Voice Transmissions	Microphone, connected to front panel mic input -- with transceiver in "SSB" mode (using whichever sideband is desired for the band)
PACTOR digital transmissions	Rear Panel Accessory socket -- with transceiver in "SSB" mode (using upper sideband)
All "soundcard" modes including both connected modes (ARDOP, WINMOR) and broadcast modes (PSK31, FT8, MT63, Olivia, etc)	USB (Universal Serial Bus) digital signals from the computer -- with transceiver in "SSB - Digital" mode

Because both the PACTOR digital and normal voice single sideband transmissions use the same ("SSB") mode -- the ICOM 7300 is configured to automatically accept analog input from either the front panel mic caonnector OR the rear panel accessory connector -- so if you are sending PACTOR and have talking noise in the room, it may be picked up by the microphone and go out along with the PACTOR transmissions (in the CW/DATA section of the band!) -- so an important suggestion:

PHYSICALLY REMOVE THE MICROPHONE TEMPORARILY WHILE CONDUCTING PACTOR DIGITAL COMMUNICATIONS WITH THE FLORIDA BAPTIST DISASTER RELIEF ICOM 7300 TO AVOID ACCIDENTAL MICROPHONE TRANSMISSIONS

Getting the right gain/volume and other settings for the incredibly-configurable ICOM 7300 external inputs, turns out to be rather important! Before these adjustments, I was unable to get things to work out well at all for PACTOR. The solutions were found in a helpful post by Demetre Valaris SV1UY here: https://groups.io/g/pactor/topic/scs_pactor_config_with/14354355? p=,,,20,0,0,0::recentpostdate%2Fsticky,,,20,2,0,14354355.

Settings entered into the ICOM 7300 via its "MENU" and "SET" configuration menus:

ACC/USB AF Output Level	15%
ACC MOD Level	15%
DATA MOD	ACC

It is important to have adequate filter bandwidth for both the PACTOR (using SSB) and for the soundcard modes (which are set to use USB-D) -- do not change FIL1 (FILTER ONE) to anything narrower than 2400 Hz. Adjusting that filter setting takes a bit of getting used to, and recommendation is not to adjust further if it is set properly.

WINLINK WINMOR OR ARDOP **"RADIO SETUP"**

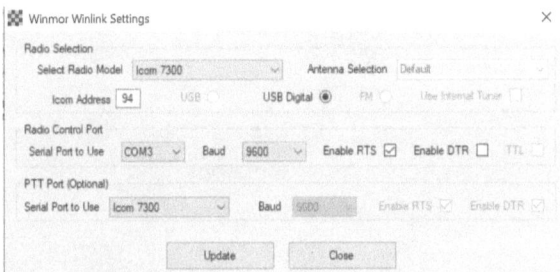

(ARDOP is similar) **Note that USB Digital is selected.** This allows the computer to control to send and receive via the USB (universal serial bus) connection.

WINLINK PACTOR "**RADIO SETUP**"

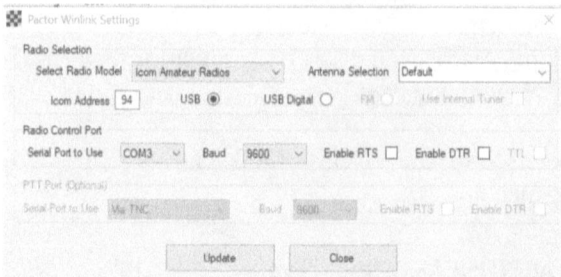

Note that "USB" rather than the USB-Digital is selected -- this allows the external PACTOR modem to send and receive signals via the analog cable to the rear panel ACCESSORY connection.

For proper linear (non-distorted) operation of the transmitter, the gain settings for the audio signals to be transmitted should be adjusted so that the signals don't cross into the "red" portion of the Power Output scale (or little to no ALC action if this is monitored). This has been preset for the WINMOR/ARDOP USB-Digital. For the PACTOR, the transmission levels of the individual Pactor modem are set as shown in the following figure:

WINLINK PACTOR TNC SETTINGS:

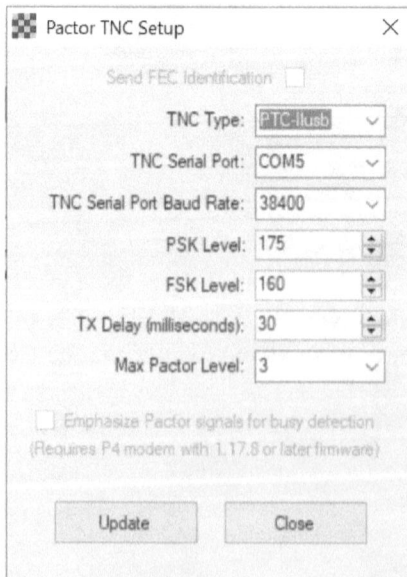

The proper com port for the Bluetooth connection from the PACTOR modem could possibly change and requires a bit of experimentation and examination of the WINDOWS SETTINGS "Device Manager" Com-Ports displays to figure out.   Bluetooth connections from Pactor Modems always TWO sequential com ports.   The one to select in the WINLINK tnc setup is the higher, or ODD number.   Leave the TX Delay at 30 milliseconds, and for USA operation, the max pactor level is 3.   The Serial port Baud Rate must be set to 38400 (the rate the Pactor Modem prefers)

APPENDIX TWO
EXPORTING AND IMPORTING
WINLINK MESSAGES

A strength of many disaster ministry organizations is they have planned on truly MASSIVE operations for feeding thousands of disaster survivors.

Planning for outbound so-called "health & welfare" communications from disaster areas to let loved ones know the status of survivors, has to be approached with the same planning for truly large numbers of messages. Just planning to "use the comms trailer computer to type them in" probably isn't adequate planning. One person can probably read a hand-scribbled message and type it into WINLINK in a minute or two, allowing for a throughput of perhaps 30 messages per hour. If you just received 300 messages by return delivery from several shelters, you could be looking at 10 hours of typing on one locked-down computer.

Luckily WINLINK provides a system allowing division of labor -- which can be replicated into as many windows-based computers as you can muster.

WINLINK allows for XML EXPORT of one or even a batch of messages -- and then IMPORT into the actual computer doing the radio transmission.

Let's look into how this work, and how it can be integrated into a significant ministry service to large numbers of persons.

PHOTOCOPY. Have a large number of copies of blank, self-explanatory message forms that can be delivered directly to disaster survivors, or by delivery (perhaps with ready-to-serve food) to shelters being managed by other service organizations. Consider using a form such as that on page 23 of this text.

COLLECT. Arrange for the orderly collection of responses and return to your disaster ministry, so that none are lost.

PREPARE TYPISTS AND COMPUTERS. Get WINLINK installed on several typists' computers. Messages will "go out" under the callsign chosen on the instance of WINLINK into which they were created -- so install the desired callsign, etc., on each typist's computer. They can all be the same callsign, or they can be different -- doesn't really matter.

Have the typists go to work entering all the messages. It is simplest just to have them "post out outbox" even though the messages won't be sent from that computer

TRANSFER FROM TYPISTS' COMPUTER TO RADIO COMPUTER:

1. Select the OUTBOX of a typist's computer.

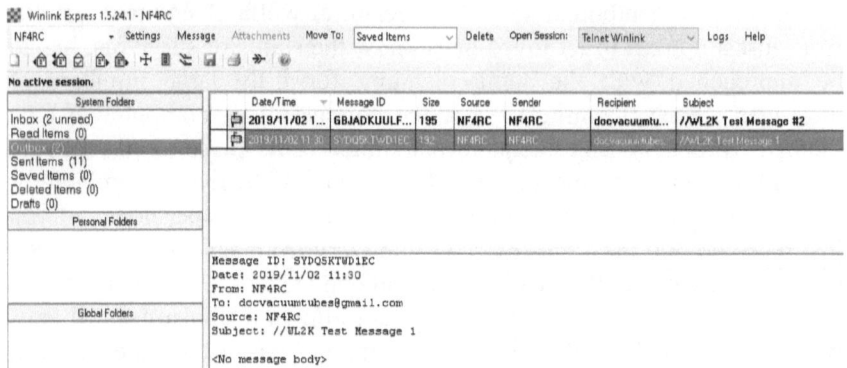

2. Select ALL the messages that you want to export (click on the top one, then SHIFT-CLICK on the bottom one, for example).

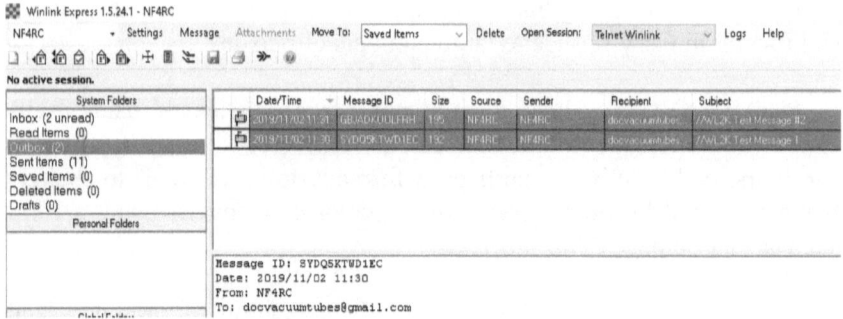

3. Select **Message | Export messages**

4. When offered, BROWSE to find the location of your transport location (**might be a simple USB thumb drive**, or a networked directory if you have your computers on a local area network). Export the files.

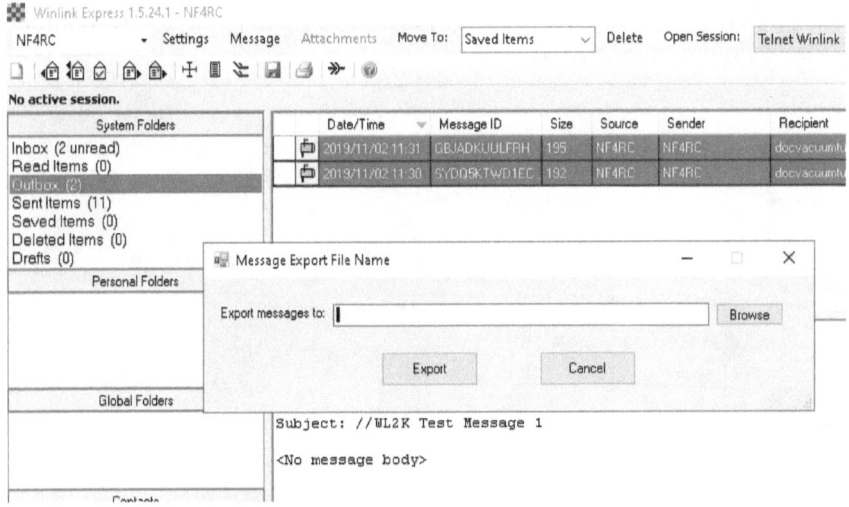

When successful, you'll get a notice like this:

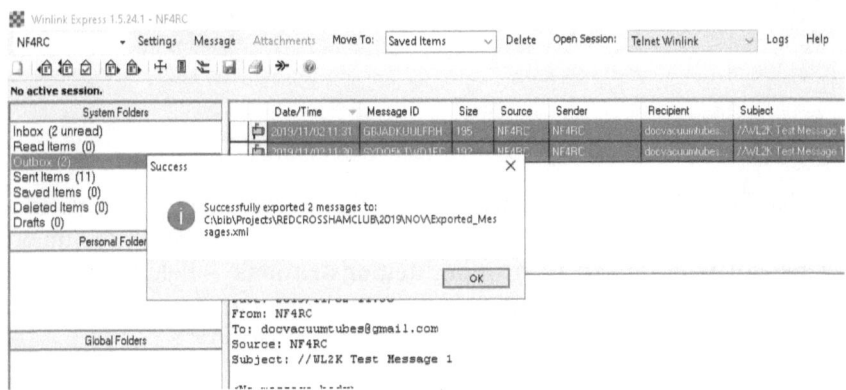

AT THE RADIO COMPUTER

5. Move to the actual computer that will do the WINLINK connection.

6. Select **Message | Import Messages.**

7. Find the location where the messages were exported to, select the XML file, and import -- the messages will automatically go into your

OUTBOX, but to be safe, it might be wise to already have selected that System Folder in WINLINK.

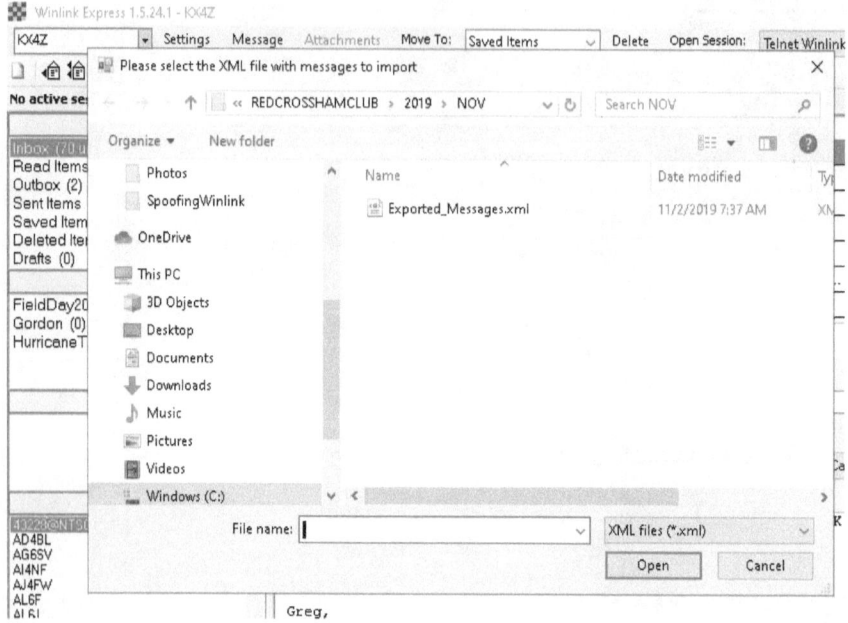

NOTE FOR TRAINERS: If those messages already exist ANYWHERE in that WINLINK system, they don't seem to import. **This can trip you up in training demonstrations** -- delete them even from the "deleted" folder if you want to re-import them as part of a demo!

NOTE FOR OPERATORS: If the messages were created under a different WINLINK callsign, you won't be able to edit them -- but you can still send them.

8. When the messages import successfully into the radio-connected computer, you'll get a helpful notice telling you how many you imported.

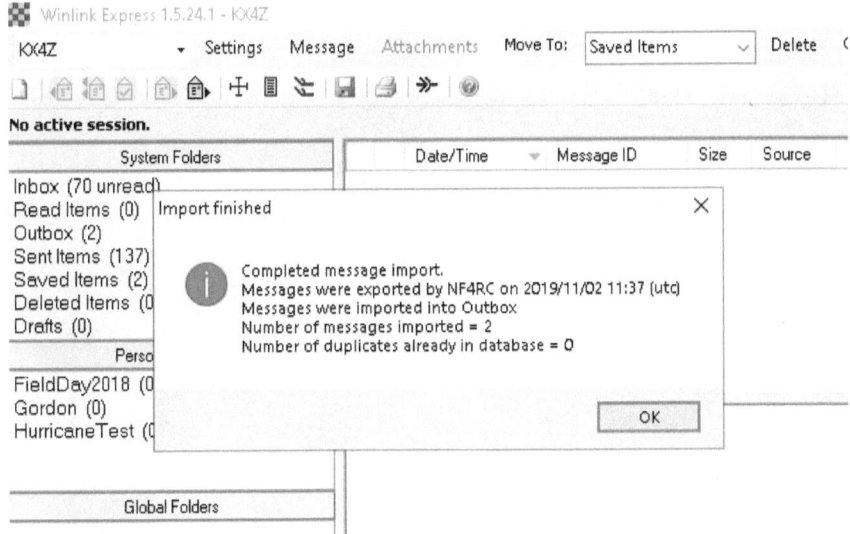

9. Initiate a radio session (e.g., WINMOR, ARDOP or PACTOR) make a connection to an RMS station, and all your messages will transfer out of the disaster area. If you lose your connection at any point, messages that were not fully and correctly transferred will still be there in your outbox. Make another connection and the transfers will resume. A little practice at this is helpful!

THANKS for your ministry to others!

APPENDIX THREE
ADDITIONAL REFERENCE INFORMATION

SUGGESTED DEPLOYMENT CHECKLIST

In addition to your normal deployment items for clothing, medications, etc, consider the following:

	FCC Frequency Chart from ARRL See: http://www.arrl.org/graphical-frequency-allocations You can also get these in recent issues of QST. I would suggest taking SEVERAL or photocopying
	PRINT OUT Repeater Information in the area you are deploying into.
	PRINT OUT (or add to your WINLINK CONTACTS list) emails for ARES officials, ARRL officials, County Officials, Florida Baptist Disaster Relief officials, and anyone else you might need to reach from the disaster area.
	Look up the GRID LOCATOR of the area you're going toward, enter that in WINLINK SETUP, and update all your channel selection tables!!!!
	Update all your software for any relevant system
	Perform any required WINDOWS or other updates
	Put copies of important installation software on a flash drive
	HF radio gear
	Digital radio gear
	Laptop computer
	VHF/UHF radio gear, and spares if possible
	Chargers for your radios and your CELL PHONE!!!
	Extension cord
	Some connection for 12V power
	Some power supplies to run from AC power

	Consider a generator or other electrical source
	Antenna Equipment -- for as many bands as possible
	Slingshot or equivalent gear
	Antenna Tuners
	SWR meters, or antenna analyzers
	Various coax "patch cables"
	Small ground rod
	Extra wire
	Tool kit, including digital volt meter
	Lightning protection gear
	Extra fuses and any necessary repair equipment
	Soldering, crimping equipment

ALACHUA COUNTY STANDARD
RADIO CONNECTIONS RJ-45 PINOUT

RJ45 Pin	Signal
1	Mic
2	Ground
3	Push to Talk
5	Speaker audio

ICOM CI-V NUMBERS AND BAUD RATES FOR COMMON RADIOS (AUTO ACCEPTS 1200 / 9600 OR 19200)

ICOM RADIO	CI-V NUMBER	BAUD
706 MK II	4E	AUTO
706 MK IIG	58	AUTO
718	5E	AUTO
725	28	1200
728	38	1200
746PRO	56	AUTO
7000	70	AUTO
7200	76	AUTO
7300	94	AUTO
7600	7A	AUTO
7610	7A	AUTO

For more information on CI-V numbers see: http://www.docksideradio.com/Icom Radio Hex Addresses.htm

Common Grid Squares for Florida

City	Grid Square	Latitude	Longitude
Avon Park	EL97fo	27.5959 / 27° 35' 45" N	-81.5062 / 81° 30' 22" W
Cape Coral	EL96an	26.5629 / 26° 33' 46" N	-81.9495 / 81° 56' 58" W
Daytona	EL99lf	29.2108147 / 29° 12' 38" N	-81.0228331 / 81° 1' 22" W
Deland	EL99ia	29.0283 / 29° 1' 41" N	-81.3031 / 81° 18' 11" W
Eustis	EL98du	28.8528 / 28° 51' 10" N	-81.6854 / 81° 41' 7" W
Gainesville	EL89up	29.6516 / 29° 39' 5" N	-82.3248 / 82° 19' 29" W
Hialeah	EL95uu	25.8576 / 25° 51' 27" N	-80.2781 / 80° 16' 41" W
Inverness	EL88uu	28.8355 / 28° 50' 7" N	-82.3314 / 82° 19' 53" W
Jacksonville	EM90eh	30.3322 / 30° 19' 55" N	-81.6556 / 81° 39' 20" W
Key West	EL94cn	24.5551 / 24° 33' 18" N	-81.78 / 81° 46' 48" W
Miami	EL95vs	25.789 / 25° 47' 20" N	-80.2264 / 80° 13' 35" W
Milton	EM60lp	30.6324 / 30° 37' 56" N	-87.0397 / 87° 2' 22" W
Ocala	EL89we	29.1872 / 29° 11' 13" N	-82.1401 / 82° 8' 24" W
Orlando	EL98hm	28.5383 / 28° 32' 17" N	-81.3792 / 81° 22' 45" W
Panama City	EM70ed	30.1588 / 30° 9' 31" N	-85.6602 / 85° 39' 36" W
Pensacola	EM60jk	30.4213 / 30° 25' 16" N	-87.2169 / 87° 13' 0" W
Perry	EM80fc	30.1174 / 30° 7' 2" N	-83.5818 / 83° 34' 54" W
St Augustine	EL99iv	29.8947 / 29° 53' 40" N	-81.3144 / 81° 18' 51" W
St Petersburg	EL87qs	27.7731 / 27° 46' 23" N	-82.64 / 82° 38' 24" W
Tallahassee	EM70uk	30.4383 / 30° 26' 17" N	-84.2807 / 84° 16' 50" W
Tampa	EL87sw	27.9506 / 27° 57' 2" N	-82.4572 / 82° 27' 25" W

FRS/GMRS FREQUENCIES

FRS CHANNEL	GMRS CHANNEL	FREQUENCY (MHZ)
1	9	462.5625
2	10	462.5875
3	11	462.6125
4	12	462.6375
5	13	462.6625
6	14	462.6875
7	15	462.7125
8		467.5625
9		467.5875
10		467.6125
11		467.6375
12		467.6625
13		467.6875
14		467.7125
15	1	462.5500
16	2	462.5750
17	3	462.6000
18	4	462.6250
19	5	462.6500
20	6	462.6750
21	7	462.7000
22	8	462.7250

MICROPHONE JACK PINOUT AS VIEWED FROM THE EXTERIOR OF THE TRANSCEIVER

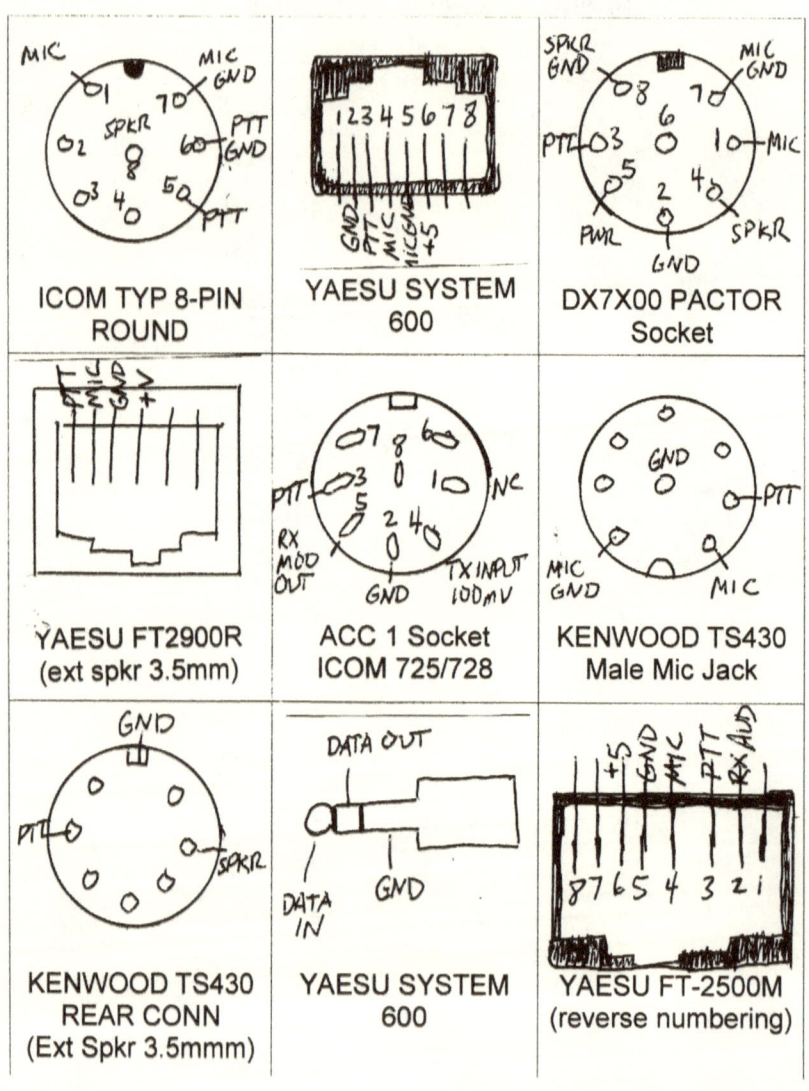

ICOM TYP 8-PIN ROUND	YAESU SYSTEM 600	DX7X00 PACTOR Socket
YAESU FT2900R (ext spkr 3.5mm)	ACC 1 Socket ICOM 725/728	KENWOOD TS430 Male Mic Jack
KENWOOD TS430 REAR CONN (Ext Spkr 3.5mmm)	YAESU SYSTEM 600	YAESU FT-2500M (reverse numbering)

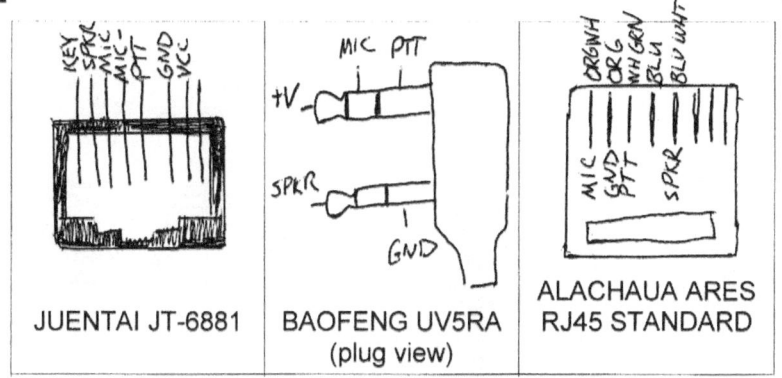

| JUENTAI JT-6881 | BAOFENG UV5RA (plug view) | ALACHAUA ARES RJ45 STANDARD |

Radio Horizon = approx 1.4 * $\sqrt{\text{antenna height}}$
where Radio Horizon is in miles, and antenna height is in feet.

VHF/UHF max range (due only to curvature of earth) is sum of distance to radio horizon from each antenna at the two stations trying to connect.

Max contact due to radio horizon (in miles) for two stations:

Ant. Ht.	10ft.	20ft.	40ft.	80ft	120ft	200ft	300ft
10ft	8	10	13	17	19	24	28
20ft	10	12	15	19	21	26	30
40ft	13	15	18	22	24	29	33
80ft	17	19	22	26	28	33	37
120ft	19	21	24	28	30	35	39
200ft	24	26	29	33	35	40	44
300ft	28	30	33	37	39	44	48

Gordon L. Gibby MD KX4Z

Federal Communications Commission
Wireless Telecommunications Bureau

RADIO STATION AUTHORIZATION

LICENSEE: NORTH AMERICAN MISSION BOARD SBC			

Call Sign	File Number
WQAL495	0006250270

Radio Service
IG - Industrial/Business Pool, Conventional

ATTN: CATHY MILLER
NORTH AMERICAN MISSION BOARD SBC
4200 N POINT PARKWAY
ALPHARETTA, GA 30022-4176

Regulatory Status
PMRS

Frequency Coordination Number

FCC Registration Number (FRN): 0010464089

Grant Date	Effective Date	Expiration Date	Print Date
05-06-2014	05-06-2014	06-28-2024	05-07-2014

STATION TECHNICAL SPECIFICATIONS

STATION TECHNICAL SPECIFICATIONS

Fixed Location Address or Mobile Area of Operation

Loc. 1   **Area of operation**
Operating Nationwide including Hawaii, Alaska, and US Territories.
Location 1 Special Condition
Area of operation is restricted to south of Line A and/or west of Line C.

Antennas

Loc No.	Ant No.	Frequencies (MHz)	Sta. Cls.	No. Units	No. Pagers	Emission Designator	Output Power (watts)	ERP (watts)	Ant. Ht./Tp meters	Ant. AAT meters	Construct Deadline Date
1	1	000151.62500000	MOI	100		11K2F3E	40.000	40.000			
1	1	000151.76000000	MOI	10		11K2F3E	35.000	35.000			
1	1	000154.52750000	MOI	100		11K2F3E	35.000	35.000			
1	1	000464.50000000	MOI	100		11K2F3E	35.000	35.000			
1	1	000464.55000000	MOI	100		11K2F3E	35.000	35.000			

(This information is publicly available for any license.)

Closer view of the frequencies:
Antennas

Loc No.	Ant No.	Frequencies (MHz)	Sta. Cls.	No. Units	No. Pagers	Emission Designator	Output Power (watts)	ERP (watts)
1	1	000151.62500000	MOI	100		11K2F3E	40.000	40.000
1	1	000151.76000000	MOI	10		11K2F3E	35.000	35.000
1	1	000154.52750000	MOI	100		11K2F3E	35.000	35.000
1	1	000464.50000000	MOI	100		11K2F3E	35.000	35.000
1	1	000464.55000000	MOI	100		11K2F3E	35.000	35.000

SHARES121013

94

HF IONOSPHERIC PROPAGATION

Your first day on deployment, every two hours you should be testing for the strongest RMS stations you can reach and filling out a table for use should it become necessary. Here is an example table for your use:

TIME (local)	BEST STATIONS/ BAND		TIME	BEST STATIONS/ BAND	
	AMATEUR	SHARES		AMATEUR	SHARES
0000			1200		
0200			1400		
0400			1600		
0600			1800		
0800			2000		
1000			2200		

Gordon L. Gibby MD

ABOUT THE AUTHOR

Gordon L. Gibby is about to retire from the active practice of Anesthesiology. Trained originally as an electrical engineer, as his career moved to the latter stages, he returned to the ham radio pursuits of his youth and discovered something completely NEW had developed in amateur radio --- "digital" communications. Initially a bit put-off by all the computers, he eventually warmed up to all this and recognized the huge advantages to emergency communications. However, he continues to maintain that the most important thing for a communications volunteer is to be very broadly experienced.

Gordon became a follower of Jesus Christ at a Baptist Revival service in the 10th grade, invited by a friend. The news of how to repent and accept Jesus' sacrificial atonement for our sins hadn't really come across in the mainline denomination Gordon had been attending all his life. Since then, there have been the usual ups and downs of life but he has persevered in the Christian Walk. He met his wonderful wife, Nancy, while both were on a short-term medical mission to Haiti. They have raised three wonderful young men, two of them adopted at birth, and also sheltered a couple of teenagers from widely disparate backgrounds. All of these different kinds of people in their live have greatly enriched Gordon and Nancy's life experience.

NOTES

NOTES

NOTES

NOTES